Risk-Based Thinking

Society at large tends to misunderstand what safety is all about. It is not just the absence of harm. When nothing bad happens over a period of time, how do you know you are safe? In reality, safety is what you and your people do moment by moment, day by day to protect assets from harm and to control the hazards inherent in your operations. This is the purpose of RISK-BASED THINKING, the key element of the six building blocks of Human and Organizational Performance (H&OP).

Generally, H&OP provides a risk-based approach to managing human performance in operations. But, specifically, RISK-BASED THINKING enables foresight and flexibility—even when surprised—to do what is necessary to protect assets from harm but also to achieve mission success despite ongoing stresses or shocks to the operation. Although you cannot prepare for every adverse scenario, you can be ready for almost anything. When RISK-BASED THINKING is integrated into the DNA of an organization's way of doing business, people will be ready for most unexpected situations. Eventually, safety becomes a core value, not a priority to be negotiated with others depending on circumstances.

This book provides a coherent perspective on what executives and line managers within operational environments need to focus on to efficiently and effectively control, learn, and adapt.

Tony Muschara is the Principal Consultant at Muschara Error Management Consulting, LLC, specializing in human error risk management in high-hazard, industrialized environments, and is a Certified Performance Technologist (CPT), awarded by the International Society for Performance Improvement (ISPI). Tony became a nuclear plant simulator instructor at Farley Nuclear Plant (1983–1985) and then joined the Institute of Nuclear Power Operations (1985–2007) where he authored several human performance guidelines for the nuclear industry. Many of his documents were adopted by the U.S. DOE and the World Association of Nuclear Operators. Tony received a BS degree in general (mechanical) engineering from the U.S. Naval Academy, served in the U.S. Submarine Service (1975–1982), Qualified in Submarines, qualified as Engineer of Naval Nuclear Propulsion Systems, and retired as Captain, USNR. Tony also received an MBA from Kennesaw State University, USA.

"This book provides sound principles for managing hazardous operations that would otherwise take years to understand through trial and error. The real-world application of Muschara's material has produced immediate results."

—Lee Reynolds, SI Group, Inc.

"The 'Must-Read' book on reducing risks from human error; Tony' s passion is for people who do hazardous work, keeping them safe, helping them succeed. Tony is one of the best, an exemplar practitioner of resilient, reliable operations."

—W.E. Carnes, U.S. Department of Energy (retired),
former Senior Advisor for Human Performance & High Reliability

"Tony's book brings together much of the new thinking about Human Error and Safety in a practical guide to those who manage any organization, whether it be safety-critical or simply an economic enterprise. Very helpful and easy to follow."

—John Wreathall, Independent consultant in organizational safety for
nuclear power, health care and transportation

"Tony Muschara begins to build a much-needed bridge from a traditional view of human performance to the 'new view', moving toward increasing resilience and reliability with an understanding that work is variable thus adaptability is key to success."

—Elizabeth Lay, founder and principal,
Applied Resilience, LLC, Houston, Texas

Risk-Based Thinking

Managing the Uncertainty of
Human Error in Operations

Tony Muschara

Routledge
Taylor & Francis Group

LONDON AND NEW YORK

First published 2018
by Routledge
2 Park Square, Milton Park, Abingdon, Oxon OX14 4RN

and by Routledge
711 Third Avenue, New York, NY 10017

Routledge is an imprint of the Taylor & Francis Group, an informa business

British Library Cataloguing-in-Publication Data
A catalogue record for this book is available from the British Library

Library of Congress Cataloging-in-Publication Data
A catalog record for this book has been requested

ISBN: 978–1–138–30247–1 (hbk)
ISBN: 978–1–138–30249–5 (pbk)
ISBN: 978–0–203–73173–4 (ebk)

Typeset in Bembo
by HWA Text and Data Management, London

Contents

Figures

Tables

Special thanks

First and foremost, I am greatly indebted to Dr. James Reason, retired Professor Emeritus of Psychology at the University of Manchester, England. Much of the content of this book is based on his publications. He has been a consultant to numerous organizations throughout the world, a frequent keynote speaker at international conferences, and is the author of several renowned books including *Human Error* (1990) and *Managing the Risks of Organizational Accidents* (1997). His principal research area was human error and the way people and organizational processes contribute to the breakdown of complex, well-defended technologies such as commercial aviation, nuclear power electric generation, process industries, railways, marine operations, financial services, and healthcare. His body of work has inspired me throughout my career. Today, I still view *Human Error* as my primary reference for the fundamentals of human performance. To me *MROA* is the original and essential source for understanding organizational influences on the occurrence of serious events. Thank you, Jim.

During the years of research and writing this book, many people gave generously of their time and shared valuable insights on safety and human performance with me. Many read earlier drafts, offering sage advice that more than once proved crucial to the clarity of the final manuscript. The clients, colleagues, and friends listed below (alphabetically) offered very perceptive and occasionally very frank and penetrating criticism. I owe them all a great debt of gratitude.

Brian Baskette—Facilitator, trainer, evaluator, and Principal Program Manager for Human Performance at the Institute of Nuclear Power Operations (INPO); colleague at International Society for Performance Improvement (ISPI); and friend.

Dr. Carl Binder—Widely published author, keynote speaker, and SixBoxes® performance improvement consultant; studied human performance with B.F. Skinner at Harvard University.

Mike Blevins—Management consultant; former plan manager of Comanche Peak nuclear power plant and retired chief operating officer for Luminant.

Bill Blunt—Performance consultant; retired Global Human Performance for Amgen, Inc.; retired U.S. Navy Submarine Master Chief; and friend.

Cathy Brown—Performance consultant; Instructional Architect at INPO; and colleague at ISPI.

Dr. Willian Brown—Human factors scientist at U.S. Department of Energy's (DOE) Brookhaven National Laboratory.

Randy Cadieux—Performance consultant, author, speaker, and coach; served 20 years in the U.S. Marine Corps and as a plane commander of C-130 Hercules transport aircraft.

Dr. Todd Conklin—Consultant in resilience engineering applications; former Senior Advisor at U.S. DOE's Los Alamos National Laboratory; author; Dave Fink's college roommate; speaker/standup comic; and adult Boy Scout.

Ron Farris—Human performance consultant; formerly human factors specialist at U.S. DOE's Idaho National Laboratory; served in the U.S. Navy as a nuclear operator on the USS Nimitz (CVN-68); and friend.

Dave Fink—Operations principal project manager for National Security Technologies, LLC at the U.S. DOE's Nevada National Security Site; and friend.

Geof Fountain—Human performance advisor at U.S. DOE's Savannah River National Laboratory and systems thinker; and friend.

Bill Freeman—Executive coach; management consultant; former Vice President, Manufacturing for Amoco Corporation; mentor and friend.

Rey Gonzalez—Human performance consultant; former manager of human performance at Southern California Edison and STP Nuclear Operating Company; coach extraordinaire; and friend.

Dr. Richard Hartley—Lead for Reliable System Performance for U.S. DOE's Pantex and Oak Ridge National Laboratories; Professional Engineer; and author.

Elizabeth Lay—Author, speaker, consultant, and practitioner in Resilience Engineering and High-Reliability Organizing (HRO) philosophies.

Dave Matherly—Retired senior manager at Browns Ferry Nuclear Plant, Tennessee Valley Authority; Qualified in Submarines: officer on the U.S. Navy submarine USS Henry L. Stimson (SSBN 655); and friend.

Jim Marinus—Currently, Human and Organizational Performance Professional at the U.S. DOE's Los Alamos National Laboratory; trainer, consultant, manager; Qualified in Submarines—served nine years on active duty as Machinist Mate in U.S. Navy Submarine service; and friend.

Dr. Jake Mazulewicz—Human performance consultant; trainer extraordinaire; speaker; former human performance specialist for Dominion Virginia Power; served in the Maryland Army National Guard as a combat engineer and paratrooper; and friend.

John Merrell—Human performance consultant; former colleague at Westinghouse Electric Company; and friend.

Michael Moedler—Head of Error Prevention System Deployment at Lonza Operations.

Dennis Murphy—Senior executive at Amgen, Inc.; retired U.S. Navy, served as Commanding Officer, USS Tucson (SSN 770) and as Commandant, Submarine Squadron 7.

Tom Neary—Software engineer, media consultant, marketing guru, professional engineer.

Bill Noll—Executive at Exelon, Inc. and former plant manager, Brunswick Nuclear Station.

Mark Peifer—Retired executive at American Electric Power; former chief operating officer at Ohio Valley Electric Corp; former colleague in human performance at the Institute of Nuclear Power Operations (INPO); Qualified in Submarines; U.S. Naval Academy classmate and friend.

Lee Reynolds—Senior Director, Environmental Health and Safety, Regulatory Affairs SI Group, Inc..

Wayne Rheaume—Human performance consultant; former Director, Performance Improvement and Human Performance at James A. FitzPatrick Nuclear Plant, Entergy Corp; and friend.

Bill Rigot—Human performance consultant; former Fellow Technical Advisor, Savannah River National Laboratory; and friend.

Gina Rester-Zodrow—Senior Manager for Learning and Performance, Amgen, Inc..

Dr. Alice Salway—Senior human factors specialist, Canadian Nuclear Safety Commission; former ergonomics and human factors consultant.

Rizwan Shah—Organizational Culture Advisor, U.S. DOE; former Chief Operations Officer for a U.S. Army combat aviation brigade.

Dr. Ralph Soule—Social scientist; researcher in High Reliability Organizing; retired Captain, U.S. Navy serving on aircraft carriers USS Dwight D. Eisenhower (CVN 69) and USS Enterprise (CVN 65).

Matt Sunseri—Entrepreneur; former President and CEO of Wolf Creek Nuclear Operating Corp.

Thomas Swanson—Retired Boeing 767 pilot with Delta Air Lines and Boeing 737 pilot Air Tran Airways; retired Commander, U.S. Navy Reserve; and best friend.

Dave Thomas—Consultant in high-reliability organizing (HRO) applications; former wildland fire manager for U.S. Forest Service.

Pierre Tremblay—President, Canadian Nuclear Partners; retired Chief Nuclear Officer, Ontario Power Generation; Chairman of the Board, Durham College.

Dr. Amy Wilson—Director, Global Biologics Manufacturing & Technical Operations Performance Development for Biogen, Inc.; devoted mother.

Dee Woodhull—Safety and management consulting; occupational safety and health networking.

John Wreathall—Consultant, author, researcher, risk manager, and friend.

In particular, I want to thank Greg Wilson, who did the heavy lifting, compiling the very first draft of the book, getting me off the dime. Also, Anne Alexander, Caitlyn Murphy, and Christina Breston exercised superb editorial skills, assisting me in the manuscript's review, revision, and proofreading during the book's intermediate and final stages of preparation.

Finally, I thank Pam, my best friend and encourager.

Tony Muschara
May 2017

Introduction

Human error is the thief of human happiness and the slayer of dreams, careers, potential, and all too frequently – life itself. Viewing it as anything less hostile is to willfully expose your throat to the knife.

Tony Kern[1]

The hallmark of an HRO [high-reliability organization] is not that it is error-free but that errors do not disable it.

Karl Weick and Kathleen Sutcliffe[2]

What this book is about

Every aspect of commercial and industrial activity involves people. To compete, much more, survive, in the marketplace, an organization must create enduring value for its customers without suffering serious harm to its employees, its communities, or its products and property. An organization must accomplish its daily operations through the hands-on work of people. The term "operations" refers to the application of some physical means to produce and/or deliver a product that achieves an organization's mission.[3] Production means work, and work means risk—risk of harm to things of value to individuals, the organization, and to society. Human error, and the harm it can trigger, creates a risk to the organization and its assets. It is an inescapable fact that we as human beings possess the prospect to err—it's not a matter of "if," but "when." Human error is pervasive and to some extent unpredictable—introducing uncertainty into industrial operations with the ever-present specter of harm.

Error is the raw material of quality problems, waste, accidents, and catastrophes—industry often sets goals of "zero errors." Society is for the most part ashamed of error. It is often the spotlight of public ridicule. It is usually considered the opposite of good performance. Also, it's not uncommon for the number of errors to be used often as a measure of bad or unprofessional performance.

Human error in operations cannot be left to chance—the risk must be managed. As Dr. James Reason has said numerous times, "Wherever there are human beings there is human error." The risk of human error is a chronic residual risk that requires ongoing attention, and you are responsible for managing it, which is the purpose of this book. The management of risk requires that you seek, by various means, to control your exposure to risk and the consequences of human error. The scope of this book is limited to events of the human kind, not those spawned by technical failures (equipment and software) or natural causes. That implies your need for a system of thought—a *way of thinking* about human and organizational performance that will aid you in your attempts to manage the risk.[4]

There is a tendency to misunderstand what safety is all about. It is not just the absence of harm.

When nothing bad happens over a period of time, how do you know you are safe? In reality, safety is what you and your people do moment by moment, day by day, to protect your assets from harm and to control the hazards intrinsic in your operations. This is the purpose of RISK-BASED THINKING, the essential and core function of the six building blocks of Human and Organizational Performance (H&OP). As a system of thought, H&OP provides a risk-based approach to managing human performance in operations. Specifically, RISK-BASED THINKING enables foresight and flexibility—even when surprised—to do what is necessary to avoid harm to assets but also achieve mission success despite ongoing stresses or shocks to the organization. *Although you cannot anticipate every possible harmful scenario, you can be ready for almost anything.* A bold assertion, but I believe it. When RISK BASED THINKING is integrated into the DNA of an organization's way of doing business, people will be ready for most unexpected situations. Eventually, people in your organization will accept safety as a core value, not as a priority to be negotiated with others depending on circumstances.[5]

"Safety should not be considered a priority, but a value with no compromise."

—E. Scott Geller
Author: *The Psychology of Safety* (1998)

Traditionally, managers have attempted to control human performance by charging the members of the workforce to avoid human error—to pay attention. In contrast to this mistaken and ill-advised strategy, I offer an alternative approach to managing human performance—*manage the risk human error poses to your assets during operations*. Strategically, this approach is best characterized as:

- *Flexible:* enhancing the mindfulness of front-line personnel—those who work directly with intrinsic operational hazards—granting them the flexibility to *adjust* to the workplace—to do what is necessary to protect assets from harm, when procedures or expectations are unclear. Workplace risks are not static but dynamic, varying during the production work day and during jobs, some occurring unexpectedly.
- *Risk-based:* identifying and applying *controls* that target the most frequent operations with the greatest potential for harm to key assets—ensuring the right things go right the first time, every time—during transfers of energy,

movements of mass, and transmissions of information that could cause harm if control is lost.

- *Systemic:* understanding how organizational systems influence (1) behavior choices and (2) the presence and effectiveness of defenses. Individual human performance (**Hu**) and safety, using the broad sense of the term, in organizations are optimized from a systems perspective.[6] A variety of organizational functions create avenues for the "flow of influence" on **Hu** and on the integrity of assets and defenses—outcomes of system alignment. *Learning* is key to proper system alignment for both safety and productivity.

At the risk of oversimplifying the task of managing **Hu** and its risk, I describe in this book six key functions that I strongly believe are necessary to manage not only workplace **Hu** but also the system within which **Hu** occurs. This book explores these risk-based functions, which form the building blocks of H&OP:

1 RISK-BASED THINKING.[7] As an *operating philosophy*, RISK-BASED THINKING augments the capacity of front-line workers to recognize changing operational risks and to adjust their behaviors appropriately. It is not intended to be a program, but a way of thinking. Additionally, people possess a deep-rooted respect for the technology and its intrinsic hazards as well as an on-going mindfulness of impending transfers of energy, movements of mass, and transmissions of information.

2 CRITICAL STEPS. Identify procedure steps or human actions that could trigger harm to assets, if a loss of control occurs. Then, incorporate means to exercise positive control of high-risk human actions involving impending transfers of energy, movements of mass, or transmissions of information between specific hazards and specific assets.

3 SYSTEMS LEARNING. This involves the ongoing identification and correction of system-level weaknesses in an organization that (1) adversely influence behavior choices of operational personnel in the workplace and (2) tend to diminish the effectiveness of built-in defenses or inhibit the organization's resilience to the unexpected.

4 *Training and expertise.* The systematic development of the workforce's understanding of the operation's technology and a deep-rooted respect for its intrinsic hazards, which RISK-BASED THINKING depends on.

5 *Observation and feedback.* An essential form of SYSTEMS LEARNING, the real-time monitoring of operations by line managers and supervisors to see firsthand what front-line workers actually do—the choices they are making on the shop floor, (a) receiving feedback on how management systems influence front-line worker performance, and (b) giving feedback to front-line workers on their work practices.

6 *Integration and execution.* The systematic and disciplined management of risks associated with H&OP (a management function), while enabling RISK-BASED THINKING and a chronic uneasiness in all facets of operations.

The first three building blocks form the core risk-management functions of H&OP, while the last three—more management-oriented—support the first three. A more in-depth description of the building blocks is provided in Chapter 2.

This book is *not* about achieving excellence or flawless performance, improving **Hu**, or even preventing human error, although these will be natural outcomes of the ideas found herein. This book is *not* intended as an all-encompassing guide to managing all forms of human performance. There are numerous publications devoted to human factors engineering, lean manufacturing, Six Sigma, operational excellence, process safety, and "inherently safer design." These programs provide enterprise-wide methods of optimizing organizational performance, but usually do not address how to manage human error risk in operations. H&OP and RISK-BASED THINKING are complementary, not contradictory, to these programs. This book limits its scope to address current-day workplace **Hu** in organizations and its operations with existing facilities, in the here and now, and will not address these other risk-management options.

Hu and H&OP

Dr. Aubrey Daniels, an authority on human behavior in the workplace, defines human performance as one or more behaviors (B), directed toward accomplishing some work output or result (R).[8] To perform is to act in a way that achieves a desired work output. Together, one's actions and the results they accomplish define one's performance, whether purposeful or in error.[9] Preferably, from a business perspective, the value added by the work output (results) should exceed the cost of the behavior used to produce them.[10] Whenever the abbreviation, **Hu**, is used in this book, it refers to individual **Hu** in contrast to its collective perspective. Often, I will refer to **Hu** in relation to the broader context of the organization within which people perform. To distinguish individual human performance, **Hu**, from human and organizational performance, I use the short form, H&OP.[11] The concept of H&OP refers to the collective performance of an organizational unit involving the work outputs of many workers performing within the systemic context of the organization's technical and social environments for the purposes of preserving the safety of assets and creating resilience. Eventually, you will see H&OP as a systematic and systemic management process aimed at controlling the variability of **Hu** in the workplace.

Hu = B + R

In the game of baseball, the "battery" serves as the team's core defensive effort. The battery includes the pitcher and the catcher. Figuratively, the battery represents **Hu**. If the battery does its job well, the other seven members of the team are unnecessary. But this is rarely the case. In every game, all eyes are on the pitcher, the catcher, and the batter. Without a good pitcher, baseball teams do not compete. But the infield and the outfield players provide defense-in-depth should the battery "fail," which is a "common" occurrence in an exciting contest. Collectively, the team, its coaches and manager, baseball equipment, the field of play, the ballpark, the rules, the umpires, the fans, and even the weather all represent the context of H&OP.

Similarly, work by an individual (**Hu**) and the object of the individual's focus (an asset and its intrinsic hazards) form the battery of H&OP. This is where work is done. Employees are hired to do something. Creating value means risk—workers touching things. When people work, they come into close, physical contact with the organization's assets, raw materials, products, and services, as well as the intrinsic operational hazards in the workplace. It is at this interface where humans, in their effort to add value, manifest risk at the same time. This is where human error can be detrimental to safety, reliability, quality, and production. This is why work in the workplace is the primary target for risk management in the H&OP approach—the nexus of H&OP. Through the building blocks of H&OP managers can systematically and systemically manage the risk introduced by **Hu** in the battery.

Common sense refers to an ability of people in a community-at-large to understand an idea or concept without special knowledge or education. How to manage individual **Hu** in the workplace, much less human and organizational performance (H&OP), is not common sense. The customary approach to managing human error singles out the individual—as if the person is the problem to be rooted out. This approach—commonly referred to as the "person model"—assumes the system is basically safe, and it must simply be protected against unreliable operators. Based on my experiences working in several industries, executives, including board members, characteristically possess the person model of **Hu**. H&OP is a paradigm shift for many of them. Their view of **Hu** typically evolves from (1) their personal experiences—rather than from study, (2) overconfidence in their organizations—because of sustained periods of success, and (3) an overly confident assessment of their own understanding of the subject.[12] Needless to say, whatever concepts of **Hu** and H&OP people have used up to this point haven't worked very well. H&OP is uncommon sense.

Principles—core beliefs espoused in this book

There is a powerful bond between people's beliefs and the actions they take.[13] Decisions and actions are guided by one's beliefs about what is necessary to achieve success or to survive. Beliefs usually take the following structure: "If *A* is true, then *B* will occur. Therefore, I will do *C*." For example: "If I get too close to the edge of a cliff, I may fall off. Therefore, I will shy away from the edge." The belief is the "If–then" statement. The rule of action is what follows the word "Therefore." People's beliefs about risk, success and failure, and how work should be done, follow this same pattern and affect what they do individually, as a team, and even as a corporation. World views and the beliefs they're based upon shape the culture and norms of an organization. Do front-line workers and managers have right beliefs about human fallibility, about risks in the workplace, about how the organization really works?

Managing H&OP and its core practice, Risk-Based Thinking, effectively depends on understanding and accepting a set of fundamental first principles—core beliefs. The building block functions of H&OP are based extensively on the principles described below. All levels of management in an organization must take to heart these principles before embarking on serious changes in their approach to managing human error risk in operations. Otherwise, H&OP and Risk-Based Thinking will not take and will fail to influence people's consciousness of the potential consequences of what people do in the workplace.

"Procedures or practices must be changed as circumstances change. Principles are constant and can be applied now and in the future. If management is imbued with these principles and accustomed to using them, it will adapt to change … If management has chosen such a course, it will lead to competent and dependable operation."

—Admiral Hyman G. Rickover, USN

In the early 1990s, the U.S. commercial nuclear industry recognized more explicitly the danger human error posed to reactor safety and embarked on a new strategy for managing **Hu**.[14] The industry adopted a set of principles designed to promote safe and reliable operations. To successfully integrate Risk-Based Thinking throughout the organization, its leadership must believe in their hearts the following set of principles—core beliefs—that guide decisions, responses, and behavior choices. Notice that principles are not rules, procedures, or even practices. In his report on the assessment of General Public Utilities (GPU) and its management competence three years after the nuclear accident at Three Mile Island Nuclear Station, Admiral Rickover declared, "Procedures or practices must be changed as circumstances change. However, principles are constant and can be applied now and in the future."[15] Taking Admiral Rickover's thoughts to heart, I propose that the following principles, when understood and embraced, will effectively guide your thinking and decisions about the management of H&OP going forward. Each chapter in this book builds upon these principles.

1   *People have dignity and inherent value as human beings.* Everyone wants to be treated with respect, fairness, and honesty, characteristics that are important to building trust and communication within any organization. People should not be treated as a liability—as objects to be controlled, but as knowledgeable and respected agents of the technical side of the organization who have its best interests at heart. Relationships are integral to open communication, and there are no laws against treating people with dignity and respect.

2   *People are fallible.* To err is human—error is normal. Fallibility is a permanent, intrinsic feature of the human condition, and this trait poses a hazard when people do work. Human fallibility can be moderated, but it cannot be eliminated. It introduces uncertainty into any human endeavor, especially in hands-on industrial work. However, people are also brilliant. They possess a wide range of capabilities and can adapt and improvise to

accommodate inadequate resources, weak training, poor tools, schedule conflicts, process shortfalls, among other workplace vulnerabilities. In order to protect assets, they can adjust what they do. Assume people will err at the most inopportune time. Then, manage the risk.

3   *People do not come to work to fail.* Most people want to do a good job—to be winners, not losers. Error is not a choice—it's unintentional. Nobody errs purposefully. Error is not sin—it's not immoral. Error tends to break things. Sin, in contrast, is selfish in nature and tends to break relationships. Reprimanding people for error serves no benefit. People are goal-oriented, and they want to be effective. They adapt to situations to achieve their goals. This means well-meaning people will take shortcuts, now and then, if the perceived benefit outweighs the perceived cost and risk. This, too, is normal. All people's actions, good and bad, are positively reinforced, usually by their immediate supervisors and by personal experiences of success (as they perceive it), which sustains their beliefs about "what works" and what does not. Sometimes perceptions are wrong.

4   *Errors are predictable and manageable.* Human error is not random—it is systematically connected to the work environment—the nature of the task and its immediate environment and the factors governing the individual's performance.[16] Despite the certainty of human error over a long period for large populations, a specific error for an individual performing a particular task at a precise time and place under certain conditions can be anticipated and avoided.[17] For example, what is the most likely error when writing a personal check on 2 January of every year? You *know* the answer. However, not *all* errors can be anticipated or prevented. This is why defenses are necessary. As your organization matures in its application of RISK-BASED THINKING, there will be less dependence on predicting the occurrence of human error and more focus on protecting assets from harm.

5   *Risk is an inherent, dynamic feature of the way an organization operates.* When work is executed, various intrinsic sources of energy, tools, and material are used to accomplish the work. Consequently, hazards exist (built in) within an organization's facilities because of its purposes. For larger, more complex organizations, risk is dynamic and lurks everywhere, and it varies as an outcome of the diverse ways an organization is designed, constructed, operated, maintained, and managed as well as its tempo of operations. Safe and resilient organizations are designed and built on the assumptions that people will err, things are not always as they seem, equipment will wear out or fail, and that not all scenarios of failure may be known before operations begin.

6   *Organizations are perfectly tuned to get the results they are getting.* People can never outperform the system that bounds and constrains them.[18] All organizations are aligned internally to influence the choices people make and the outcomes they experience—good and bad. All work is done within the context of its management systems, its technologies, and its societal, corporate, and work-group cultures. Organizations comprise multiple, complex interrelationships between people, machines, and various

management systems, and managers go to great lengths to create systems for controlling the work. But we all know there is no such thing as a perfect human, perfect system, perfect process, or perfect procedure. Once systems go into operation, they prove to be imperfect.[19]

7   *The causes of tomorrow's events exist today.* The conditions necessary for harm to occur always exist before realizing the harm. Some are transient, most are longstanding. Most of the time these conditions are hidden or latent. Latent conditions tend to accumulate everywhere within an organization and pose an ongoing threat to the safety of assets.[20] These conditions are shaped by weaknesses at the organizational and managerial level, and they manifest themselves in the workplace as faulty protective features, hidden hazards, and error traps. These system weaknesses and workplace vulnerabilities usually exist long before the unwanted consequences ever come to fruition. This means that events are organizational failures.

Why read this book?

In a 2012 speech, the Honorable Hilda Solis, U.S. Secretary of Labor, stated that, "Every day in America, 12 people go to work and never come home. Every year in America, nearly four million people suffer a workplace injury from which some may never recover."[21]

In a 2008 whitepaper, a global market analyst firm, IDC Research, analyzed the cost associated with human error in the form of "employee misunderstanding." The analysis studied the financial impact on a combination of 400 U.K. and U.S. businesses with more than 5,000 employees.[22] The whitepaper defined employee misunderstanding as actions by employees who misunderstood or misinterpreted company policies, business processes, job functions—or a combination of the three. It stated the following conclusions:

- Twenty-three percent (23%) of employees do not understand at least one critical aspect of their job.
- The estimated overall annual cost to US and UK businesses is $37 billion.
- On average, businesses with 100,000 employees are each losing $62.4 million per year, equating to approximately $624 per employee.

In 1999, the U.S. Institute of Medicine's study on medical error (*To Err is Human: Building a Safer Health System*) shocked the nation with its conclusion that up to 98,000 preventable deaths occur annually as a result of medical errors in U.S. hospitals. However, as recently as April 2016, the BMJ, the *British Medical Journal*, after conducting a more rigorous study of deaths due to medical error, reported that estimates have escalated to greater than 251,000 per year, now the third leading cause of death in the U.S. behind heart disease and cancer.[23]

Recent research of European enterprises substantiated something I believe the U.S. commercial nuclear industry realized in the early 1990s.[24] Safety is a profit multiplier—it sustainably improves profits by reducing the per unit costs of production by avoiding the costs associated with unwanted events and other inefficiencies that are triggered by critical errors. Together, the IDC whitepaper and the BMJ study support the assertions by Van Dyck (et al.) that companies with a so-called "error management culture," where the people of an organization attempt to manage the risk of human error, enhanced their profitability over and above prior company performance as well as exceeded the performance of those companies without an error management culture.[25]

"Lunch Atop a Skyscraper" (Figure 0.1), the photograph that also graces the front cover of this book, was published in 1932 during the final days of construction of the RCA building in Manhattan's Rockefeller Center in New York City. You may recognize the iconic image of eleven unknown ironworkers enjoying their lunch break while sitting on a steel beam near the top of the skyscraper. Their feet and fannies dangle over both sides of a steel beam more than 800 feet above the street below. They sit at the 69th story of a 70-story structure. Although this scene was likely staged for the photographer, it captured a common, everyday occurrence. The iconic photograph characterizes the risk human error poses to key assets in your operations every day.

In the photograph, these men are casually enjoying their break—from walking the high beams. They are smoking, relaxing, and joking with one

Figure 0.1 "Lunch Atop a Skyscraper." Ironworkers eat their lunches and relax atop a steel beam 800 feet above ground, at the building site of the RCA Building in Rockefeller Center in New York, Sept. 29, 1932 © Bettmann/Corbis

another as if there were no cares in the world. After a while on the job, these men became accustomed to working at dizzying heights. They became skilled, almost a second nature, at strolling along the four-inch-wide beams. They became comfortable with their surroundings as if they were sitting in their favorite chairs at home. These are regular guys, not unlike today's front-line workers in your operation.

On average, during the building's construction, one death occurred for every ten floors built. Every year, roughly 2 percent of the workforce died on the job, and 2 percent were injured in some fashion. They were literally working on the edge. You might ask yourself, "Were these men fools or heroes? Were they out of their minds?" However, this photograph was taken during the Great Depression, and work was hard to come by. These men were desperate to earn a living to support their families. Yes, skyscraper construction was dangerous in those days. But, despite the risks, men competed strongly for whatever jobs were available. Today, they would wear a safety harness tethered to the steel, hardhats, and steel-toed shoes. Lifelines and safety nets would be installed. Today, people still work on the edge, but we are required to think differently about **Hu** and the work people do.

Today, we live in a world of high risk that, thanks to modern media, we are reminded of daily. Why is the risk so high these days when innovation and advancement are beyond what we once could imagine? Because, despite these innovations and advancements, the technologies that surround us, personally and professionally, are more complex and hazardous than ever before. Greater amounts of energy and mass are marshalled together in the same place and time. Intrinsically, people are fallible. The ironworkers were no different. Interestingly, the people who designed, built, managed, operated, and maintained these technologies haven't themselves changed all that much. Human nature is pretty stable. However, human error is not the "cause" of most of your troubles, much less your events. You'll discover as you read this book that the causes of your serious incidents reside in your systems, not so much in your people.

Who this book is written for

The lessons associated with human fallibility and organizational shortcomings are not well understood. Organizations are formed and run by human beings and, thus, are fallible just as front-line workers are. Regardless of the advances in human factors, cognitive sciences, human performance, industrial and organizational psychology, and quality assurance, human error remains one of the greatest sources of variation. Serious events continue to occur, and organizational failures continue to plague businesses and industries. Much has been written about high reliability organizing, resilience engineering, quality, and operational excellence, but little has been published from a manager's point of view about what to do, specifically, to manage the risk that human error poses to everyday operations.

Commercial organizations exist for many reasons—none of which are for safety. Companies are not formed for the sole purpose of being safe.[26] When you look at the amount of time managers spend on safety compared with productivity, safety does not appear to be number one. If safety were the definitive purpose of the organization, it would be better not to start anything at all—just stay at home and out of harm's way. Profitable outputs—products and services that society wants—are the real purpose of any enterprise, and they are, out of necessity, the organization's priority. Front-line workers are hired to produce or aid in production. Line and senior managers are tasked with creating value for an organization's customers, and they must do so within a variety of constraints, whether economic, environmental, regulatory, or social. These constraints frequently create internal conflicts during industrial operations—to operate faster, better, cheaper, and safer, all at the same time. But to survive and sustain effectiveness and profitability, the people that make up your organization must value safety, protecting assets, and controlling intrinsic hazards, in addition to accomplishing the mission. Otherwise, you're out of business—eventually. Therefore, managers must manage safety simultaneously with their production activities—safety is not a collateral duty; it is core business. This book provides executives, senior managers, and line managers responsible for the day-to-day field operation of high-hazard facilities with a logical, yet efficient approach for managing the ever-present risk of human error.

As you read the book

If you are a student of human and organizational performance, you no doubt have encountered the concepts of resilience engineering, high reliability organizing, the *New View*, Safety-II, among others. These emerging perspectives on safety and reliability have been enlightening to executives, entrepreneurs, line managers, practitioners, and consultants alike. These world views[27] of human and organizational performance encourage us to better understand how systems create the conditions for success as well as failure. I attempt to integrate *New View* principles (including those introduced earlier) into the technology offered in this book, but I do not throw away what has worked in the past. I do not espouse a total rejection of the *Old View*—some things are still worthwhile.[28]

The *Old View* (Safety-I) generally promotes an avoidance world view—tending to be reactive to problems, seeing people as the primary liability in an otherwise safe system. The *New View* (Safety-II) focuses on success—emphasizing an ongoing proactive stance against events, seeing people as the primary resource for adaptability and flexibility in a hazardous workplace.[29] As you read you'll notice an overall emphasis on adaptability and flexibility to adjust to dynamic work situations, many of which had not been foreseen by the authors of technical procedures. However, I still emphasize the fundamental importance of following procedures, avoiding unsafe situations, preventing error, and building and sustaining defenses. But things are not always as they seem. Hence, I promote a risk-based approach to **Hu** in operations that promote adherence to expectations but allow for adjustments—safe behavior choices—in the field to protect assets from harm and still achieve your economic purposes safely.

Finally, at the conclusion of each chapter, I offer some practical suggestions for implementing the ideas of that chapter—Things You Can Do Tomorrow. Most of the items listed are straight-forward ideas that can be implemented with minimal upfront planning and resources.

As with any new technology, there are new terms and phrases introduced that may not be self-evident of what they mean. A glossary is provided at the back of this book as a ready reference to clarify your understanding of unfamiliar terms and phrases as you encounter them in the text.

H&OP is a fascinating and challenging field of study for me, and never more so than when it possesses practical worth for managers of high-risk operations. I hope this book stimulates interest in how to more effectively manage the uncertainty of human error and its associated risks in organizations and their operations.

Notes

1 Kern, T. (2009). *Blue Threat: Why to Err is Inhuman*. Monument, CO: Pygmy Books (p.1).
2 Weick, K. and Sutcliffe, K. (2015). *Managing the Unexpected* (3rd edn). Hoboken, NJ: Wiley (p.12).
3 Mitchell, J. (2015). *Operational Excellence: Journey to Creating Sustainable Value*. Hoboken, NJ: Wiley (p.7).
4 Johnston, A.N. (1996). Blame, Punishment and Risk Management, in Hood, C. and Jones, D. *Accident and Design*. Abingdon: Routledge (p.72).
5 Geller, E. (1998). *The Psychology of Safety*. Boca Raton, FL: CRC Press (p.367).
6 I apply "systems thinking" very narrowly to the performance of front-line workers, who do the work of the organization and who come into direct contract with the organization's most valued assets and hazardous processes.
7 Since the first three building blocks form the core risk-management functions of H&OP, they are denoted throughout the book with a SMALL CAPS font style.
8 Daniels, A. (1989). *Performance Management: Improving Quality Productivity through Positive Reinforcement* (3rd edn). Tucker, GA: Performance Management Publications (p.13).
9 Attributed to Dr. Fred Nichols, CPT. Retrieved from http://www.performancexpress. org/2015/11/knowledge-workers-solution-paths-getting-from-here-to-there/
10 Gilbert, T. (1978). *Human Competence: Engineering Worthy Performance*. New York: McGraw-Hill (pp.7–8).
11 I attribute the combination of the terms *human* and *organizational* to the work of Dr. Najmedin Meshkati, University of Southern California, related to his study of the nuclear accidents at Chernobyl and Three Mile Island. He coined the phrase "human and organizational factors" to describe the causes of these accidents.
12 Viner, D. (2015). *Occupational Risk Control: Predicting and Preventing the Unwanted*. Farnham: Gower (p.238).
13 Connors, R. and Smith, T. (2011). *Change the Culture Change the Game*. New York: Penguin (p.67).
14 Institute of Nuclear Power Operations (1991). *An Analysis of 1990 Significant Events* (INPO 91-018), and a follow-on report from the Human Performance Initiative Special Review Committee, *Recommendations for Human Performance Improvements in the U.S. Nuclear Utility Industry* (November 1991).
15 Rickover, H. (19 November 1983). *An Assessment of the GPU Nuclear Corporation Organization and Senior Management and Its Competence to Operate TMI-1*. (p.ii).

16 Reason, J. (1990). *Human Error*. New York: Cambridge University Press (pp.4–5). Dekker, S. (2014). *The Field Guide to Understanding 'Human Error'* (3rd edn). Farnham: Ashgate (p.194).

17 Center for Chemical Process Safety. *Guidelines for Preventing Human Error in Process Safety*. American Institute of Chemical Engineers, 1994, pp.12–17, 103–107. Reason, J. (2008). *The Human Contribution: Unsafe Acts, Accidents and Heroic Recoveries*. Farnham: Ashgate (p.34).

18 Maurino, D.E., Reason, J., Johnston, N. and Lee, R.B. (1995) *Beyond Aviation Human Factors*. Aldershot: Ashgate (p.83). This principle is also supported by Rummler, G. and Brache, A. (1995). *Improving Performance: How to Manage the White Space on the Organization Chart*. San Francisco, CA: Jossey-Bass (p.64).

19 Spear, S. (2009). *The High-Velocity Edge: How Market Leaders Leverage Operational Excellence to Beat the Competition*. New York: McGraw-Hill (p.46).

20 Reason, J. (1997). *Managing the Risks of Organizational Accidents*. Aldershot: Ashgate (p.36).

21 Remarks by the Hon. Hilda Solis, U.S. Secretary of Labor, in a speech, "Workers' Memorial Day," in Los Angeles, CA. April 26, 2012. Retrieved from https://www.dol.gov/_sec/media/speeches/20120426_WMD.htm.

22 IDC Research (2008). *$37 Billion: Counting the Cost of Employee Misunderstanding*. Framingham, MA, US: IDC Research.

23 British Medical Journal (3 May 2016). Medical error—the third leading cause of death in the US. *British Medical Journal*. Retrieved from: www.bmj.com/content/353/bmj.i2139.

24 Institute of Nuclear Power Operations (4 November 1992). *Excellence Versus Cost*. A presentation by Zack Pate, President and CEO, INPO, at the INPO CEO Conference at the Stouffer Waverly Hotel in Atlanta, GA.

25 Van Dyck, C., Frese, M., Baer, M. and Sonnentag, S. (2005). Organizational Error Management Culture and Its Impact on Performance: A Two-Study Replication. *Journal of Applied Psychology*. Vol. 90. No. 6. (pp.1228–1240). DOI: 10.1037/0021-9010.90.6.

26 Agnew, J. and Daniels, A. (2010). *Safe By Accident?* Atlanta, GA: PMP (pp.33–34).

27 A *worldview* is a frame of reference that shapes a person's perceptions and interpretations of objects and experiences, as well as their daily, moment-by-moment, choices. Everything is perceived, chosen, or rejected on the basis of one's framework for thinking (known as one's *narrative* in the world of media). A worldview leads people to see what they expect and not see what they don't expect. I encourage you to read Diane Vaughn's description of NASA's worldview in her book, *The Challenger Launch Decision*.

28 Hollnagel, E. (2014). *Safety-I and Safety-II: The Past and Future of Safety Management*. Farnham: Ashgate (pp.147–148).

29 I invite you to review Table 8.1, A Comparison of Safety-I and Safety-II, in Erik Hollnagel's book, *Safety-I and Safety-II: The Past and Future of Safety Management* (2014). Farnham: Ashgate (p.147).

1 A nuclear professional

> The nuclear professional is thoroughly imbued with a great respect and sense of responsibility for the reactor core—for reactor safety—and all his decisions and actions take this unique and grave responsibility into account.
>
> Foreword of *Principles for Enhancing Professionalism of Nuclear Personnel*, Institute of Nuclear Power Operations[1]

As a twenty-something submarine officer on the *USS Andrew Jackson* (SSBN619), a fleet ballistic missile submarine, I was awed daily by the inner workings of the boat. I graduated from the U.S. Naval Academy with a degree in engineering, but that was the last time I did any "engineering." I was destined to become an operator (at least for a while). After seven years of active duty in the U.S. submarine force, I was truly an operator at heart. Eventually, I relished every aspect of "operating," from starting up to shutting down, even the daily drills to practice responding to simulated casualties. At first, that was not the case.

I was slow on the uptake—I'm a slow learner, even today. If I don't understand something, it is hard for me to function. Such was the case during my first couple of months during my skills training at a U.S. Navy Nuclear Prototype Training Unit (NPTU) near Hartford, CT. For six months, several of my classmates, enlisted personnel, and I applied the academic principles of reactor theory, thermodynamics and fluid flow, water chemistry, and health physics we had learned in a classroom at one of the Navy's Nuclear Power Schools to the practical operation of an actual land-based submarine nuclear plant. I had never before operated a nuclear reactor or a steam plant and its associated equipment. Overnight, we were literally thrown into the hazardous domains of high-pressure steam, high-voltage electricity, nuclear radiation, rotating equipment, and myriads of twisting runs of electric cables, high-temperature steam piping, pumps, tanks, and valves. It seemed everything was hot. It was a dangerous place, which became most vivid one evening when one particular Navy operating expectation nearly got me killed.

During one late-night shift in the submarine prototype's lower-level engine room, I was operating the steam plant's main feedwater and condensate systems in a training role (under the instruction of a qualified instructor). During a steam plant startup, I was directed to start up the second of two main feed pumps right away. In compliance with expectations, I intended to verify that a pump would rotate before starting its motor—the operator had to physically grab the pump shaft with both hands to spin it to ensure it would rotate freely. (I'm sure there was a rational technical reason for this practice.) Still a novice, distracted by thinking about the next steps, I had unknowingly become disoriented as to the "port" (left) and "starboard" (right) sides of the submarine's engine room. In my haste, I forgot to check the running indicator lights on the control board. At the time, the port-side pump was already running, while the second, starboard-side pump was idle. In the deafening noise of the engine room and through the earmuffs I wore for hearing protection, it wasn't obvious from the sound which pump was running. What's worse, the shaft of an operating pump looks idle even though it's rotating at 3600 RPM. I went to the wrong pump. Operators faced aft, while standing in front of the Feed and Condensate System control panel—I turned and walked right instead of left. As I reached in to rotate the pump shaft, my trainer screamed at the top of his lungs, "Stop!" Fortunately, I heard him. I was mere inches from having my arms ripped off and likely being killed. Wow!

It was my first dramatic encounter with a serious system failure, though I didn't know it at the time. I felt like a fool—I "should have" been paying attention. Looking back, I marvel at the foolishness of a policy that required a worker to handle a potentially deadly piece of equipment in order to discover whether or not it was operable. And the two checks to determine its safety—audible and visible—were nearly 100 percent unreliable given the conditions in the immediate work environment (what I call *local factors*).

I shudder to think what could have happened. It was an unsafe practice, but as a junior officer, I didn't know what to do about it. I shook it off as experience. But, I realized then, even before reporting to the real Navy, that simply following procedures do not always protect you. I was more cautious from then on.

After completing my training and qualifications at the prototype and spending a few months at "Sub School" in Groton, Connecticut, I reported to the AJ (Andrew Jackson) for my first at sea duty assignment. I remember walking down the "gang plank" in the early morning fog in Holy Loch, Scotland. Within the first hour of my arrival, without unpacking my bags, I performed a reactor startup, taking it critical in preparation for getting underway on my first submarine patrol. I actually enjoyed it—I knew what I was doing. This was an important "practical factor" for my qualification as an Engineering Officer of the Watch (EOOW). My EOOW qualification card listed knowledge requirements and several practical factors (operational evolutions) that I had to perform satisfactorily prior to qualifying to stand the watch unsupervised. Reactor startups didn't happen that often, and I had arrived just in time to take a seat at the controls of an operating nuclear reactor.

In the 1970s and 1980s, the Cold War was "raging" between the U.S. and the then Soviet Union. I had to learn not only how to operate the nuclear propulsion system, but also how to operate a submarine. My first six months or so on the submarine was devoted mostly to qualifying and developing proficiency in operating the submarine's nuclear propulsion system. Soon thereafter, my focus shifted to the forward part of the submarine to qualify in submarine operations. To be "Qualified in Submarines" meant I was knowledgeable of virtually all of the submarine's systems and technically proficient in submarine operations along with its missile and torpedo weapon systems. Within 18 months of reporting aboard, I was authorized to wear the gold dolphins submarine warfare specialty pin on the left breast of my uniform.

After qualifying in submarines, I had to become proficient in submarine warfare—this, too, fascinated me. While at sea, the officers and crew studied and practiced everything we could to reliably launch missiles, shoot torpedoes to survive attacks from the enemy, and stay afloat. We ran engineering casualty and tactics drills almost every day, except Sundays—we needed the rest. We relentlessly trained and practiced fighting, staying alive, keeping the submarine operational—ready to pull the trigger—if ordered. It never happened, thank God. The drills helped us learn what we didn't know and where we lacked skills—believe it or not, mistakes were commonplace. Once, while moored alongside, I unintentionally depressurized a reactor coolant loop below the minimum pressure allowed for the operation of a reactor coolant pump. It was like popping the top off a can of soda. Without going into the technical details, I felt horrible and terribly embarrassed for committing such a bonehead mistake. At the time, I was the most experienced engineering department officer on board, and was "qualified" as an engineer of naval nuclear propulsion systems. I wasn't supposed to make such a mistake. Several enlisted men and myself spent the next 12 hours during the night shift recovering from my mistake. They were not happy campers.

The Engineer qualification was awarded by Naval Reactors (NR) in Washington, the U.S. Navy's engineering organization for development of shipboard nuclear propulsion applications.[2] The months of in-depth study of naval nuclear propulsion systems preceding the engineer exam helped me recognize the limits and complexity of engineering design and the operation of its systems, structures, and components. After passing an eight-hour written examination and surviving several in-depth interviews by NR experts, I received a formal Navy letter that designated me as an "Engineer of Naval Nuclear Propulsion Systems." I remain proud of that accomplishment to this day. This was a pre-requisite for a subsequent assignment to a nuclear submarine as its Engineering Officer.

My experiences with nuclear submarine operations at sea instilled in me a deep-rooted respect for the ocean, nuclear power, and the complexity of the technology associated with it, and the fragility of human life and human structures. As a product of Admiral Rickover's nuclear Navy and the Cold War, I emerged from my naval career with a strong sense of what is required to operate safely and reliably—to fight and stay alive. A wary and disciplined approach to operations is forever instilled in my soul, in my subconscious mind—it has never left me, and I believe this mentality still has value today. Some call it the *Old View*. As a former nuclear navy-trained officer, I am imbued with a profound understanding of the naval nuclear principles and practices that came from being immersed in nuclear submarine operations for several years. Below, I list Admiral Rickover's time-tested industrial principles,[3] described in my own words, that I believe still apply today:

- *Respect the technology*—hazards are intrinsic in all work. All technologies possess some dynamic residual risks that cannot be eliminated. Remain mindful of the technology's potential for harm.
- *Develop expertise*—train front-line workers, including their managers, to understand the technology. The level of training and proficiency must be commensurate with the technology's complexity and risk.
- *Be conservative*—allow margin for possible unknown and unforeseen effects. Things are not always as they seem. Mistakes will be made at all levels—by executives, managers, engineers, as well as operators.
- *Know what is going on*—work out some simple and direct ways to stay informed and to have the means for an independent review. Learn—know what has happened, what is happening, and what should be changed going forward.
- *Get into the details*—when ignored, an operation could slide through recurring at-risk practices (drift away from expectation) into failure so fast that no policy decisions, however wise, could resurrect it.
- *Face the facts*—resist the temptation to minimize the potential consequences of problems and to try to solve them with limited resources, hoping that somehow things will work out. Humble yourself—ask for and accept help. Force problems up to higher levels where more resources can be applied.
- *Accept responsibility*—take ownership of all aspects of your operation. Courageously, admit mistakes. Don't pass the buck—admit personal responsibility when things go wrong.

After my first tour at sea, I went back to Sub School, not as a student, but as an instructor. I spent two years as a submarine tactics instructor, helping prepare the officers of other submarines get ready for their deployments. It was here I confirmed in my heart that I enjoyed instructing and helping people understand the what's, the how's, and the why's of what they were learning. In light of my close call at the submarine prototype and my time at sea, I realized how important expertise was in operating hazardous plant systems and equipment safely and reliably. I remember during my last couple of years on the AJ, other enlisted sailors and officers working on their own qualifications, came to me for one-on-one "checkouts" (a dialogue aimed at validating the trainee's system knowledge and understanding). To their chagrin, they sometimes found it frustrating to work with me. A signature from me on their qualification card indicated (to a third-party reviewer) that they possessed satisfactory knowledge and understanding of the system we were reviewing. However, I wasn't one to give away my signature—being "graped off."[4] Since I knew what was at stake in submarine operations during the Cold War, I did not go easy on them, and invariably, our conversations revealed critical gaps in their understanding. Far too many individuals did not recognize how dangerous a job they had and what they had to know to not only operate according to procedures, but also to ensure our survival if things did not go as planned.

At the conclusion of my tour of duty at Sub School, I resigned my commission in the U.S. Navy to pursue a career in civilian life—my wife and I had two children by this time. To me, family is more important than career. Former submarine officers were in high demand in the commercial nuclear industry in the early 1980s. Pursuing my joy of training, I joined a technical training firm as a nuclear plant control room simulator instructor. Here, I realized that things could still go wrong, even when everything was in compliance with the plant's technical specifications. You are not necessarily safe when you are "crossing a street in the crosswalk." You're still in the *"line of fire"*[5] while in the crosswalk. In any industrial operation, you still have to be alert to equipment failures, unknown conditions, and human error.

After a couple of years training nuclear plant operators, I joined the Institute of Nuclear Power Operations (INPO) in the spring of 1985, as a training evaluator.[6] Soon, I learned about INPO's Human Performance Department (HPD). Although not assigned to work there, I was immediately intrigued by its purpose and the technology—the Human Performance Enhancement System (HPES). INPO had formed the department as part of the commercial nuclear industry's response to the "President's Commission Report on the Accident at Three Mile Island in 1979." Since the report cited human error as a primary reason for the accident, HPD's goal was to help the nuclear industry avoid events triggered by **Hu** in the workplace.

I followed the work of HPD as a side interest during my first three years with INPO. During plant evaluations, I typically spent two weeks at a nuclear generating station with a team of specialists, observing work, meetings, and training activities; interviewing managers; and documenting my assessments of the quality and effectiveness of nuclear plant operations as it related to training. As the team's training evaluator, I focused on the effectiveness of training programs for licensed reactor operators as well as maintenance, engineering, chemistry, and health physics personnel. Observations of work and training heightened my interest in the arena of **Hu** and human error—I saw a lot of it. Then, in 1988, after three years at INPO, because of my keen interest, I was transferred to HPD. It changed my life.

The first real book I read on human performance was Dr. James Reason's *Human Error* (1990). It was a tough read. For me, a slow learner, it was pretty academic and abstract. Sometimes I had to read a paragraph two or three times to fully comprehend it. Regardless, I couldn't put it down. This book convinced me more than ever that I had discovered my passion. At the time, Dr. Reason was a professor at the University of Manchester in the UK, and I felt as if I had found a virtual mentor. Little did I know I would later have an opportunity to work with him.

In the early to mid-1990s, the commercial nuclear electric generating industry embarked on its initiative to better manage **Hu**. At the time, the industry did not think of **Hu** as a risk to manage, but everyone thought of it simply as a problem with front-line workers. "If they would just pay attention ...," many would say, simply assuming that the main problem was unreliable human beings in an otherwise reliable system—not an unsurprising conclusion coming from executives and line managers immersed in an engineering culture.

Dr. Reason's book helped the nuclear industry's leadership realize that an event "caused by human error" was more an organizational issue than a person issue. James Reason's insights about "latent errors and system pathogens" helped me realize that organizations, along with their technologies and associated processes, became more complex, obscuring the "real" causes of events. People did not become any more fallible during the 1980s than they were before.

An internal study by INPO in 1990 revealed that more than 70 percent of the events in the nuclear industry were "caused" by **Hu** (error).[7] Additionally, in response to several highly-publicized accidents in the 1980s, including the Chernobyl nuclear accident in 1986 and a rash of several nuclear plant incidents in the United States, INPO commissioned a "special review committee" comprising several experts in human performance and nuclear industry executives, to conduct an in-depth assessment on human performance in the commercial nuclear industry.[8] The report concluded that the nuclear industry did not fully understand human performance, especially its systemic factors, emphasizing the organizational nature of human performance—the first time **Hu** was considered more than simply "attention to detail."[9]

The industry soon realized it didn't have a way of talking about, much less managing, "human performance." It needed a vocabulary as well as an education. In order to manage anything, you have to be able to think and talk about it. Today is no different. In 1997, with the guidance of Dr. Reason, INPO published a brief handbook of guiding principles and behaviors, *Excellence in Human Performance*,[10] the first of several documents over the next decade incorporating the scientific principles of "high-reliability organizing"[11] and "resilience engineering." Shortly thereafter in early 1998, INPO developed the "Human Performance Fundamentals Course" to educate the industry on a "systems view" of human performance.

To implement the INPO training materials, my colleagues and I spent two to three years traveling the U.S., teaching and talking with management teams about **Hu**. The completed training material was used for an industry-wide train-the-trainer program, which was conducted in the late 1990s, and subsequently incorporated into nuclear electric utility training programs for front-line workers. This effort eventually expanded to more training and consultation opportunities within the nuclear power industry worldwide. Meanwhile, recognizing management's need for practical understanding of **Hu**, I wrote several documents related to human performance, one of which was a predecessor to this book, INPO's *Human Performance Reference Manual*, which was adopted by the U.S. Department of Energy and is currently available on line.[12]

Today, I pay close attention to how popular news media report industrial accidents. Usually, there are gaps in understanding—not only by those reporting but also by the managers involved. All too often the conclusion is that "bad" people cause disasters: "They should know better. After all, they're experts in that kind of work." And each time, my heart goes out to the individual who was the last one to "touch" something that triggered the harm. He or she gets the blame, even when everyone can see the "stage was set" for the given accident long before it happened.

Although the nuclear industry adopted the new perspectives of **Hu**, it was not without growing pains. Even to this day, there remains resistance among some nuclear industry managers and executives to the fundamental principles essential for effective management of **Hu** risk. Regardless of the industry, I still find that most managers and executives do not fully appreciate the systemic nature of **Hu**, focusing instead on the technical side of production—treating people as they do equipment—as components. They fail to realize that, unlike equipment, people think for themselves—they are able to adapt to local conditions as needed to achieve their goals. Similarly, managers do not always realize that safe choices emerge from the way their organization is designed, managed, and led. If managers and executives truly understood how their organizations really worked, they would not experience as many surprises as they do.

I'm not an academic—I'm an operator by training and experience. However, the outpouring of research and related publications on human performance and organizational factors in response to serious industrial accidents has been most helpful. I'm convinced the "new view" espoused by researchers and experienced practitioners in resilience engineering and high-reliability organizing (HRO) holds tremendous promise to better the lives of thousands of workers in high-hazard industries. But, it requires translation.[13] After consuming books, articles, as well as internet blogs and podcasts on resilience engineering and HRO, I look for the practical ramifications of the author's work. I attempt to translate their theories and concepts into operational principles and practices. In that sense, I hope the following chapters lead to practical, workable applications of H&OP and RISK-BASED THINKING in the workplace to protect not only what is important to your organization but also what you hold dear.

Notes

1   Institute of Nuclear Power Operations (September 1987). *Principles for Enhancing Professionalism of Nuclear Personnel*. Atlanta, GA: INPO (not public domain).
2   Duncan, F. (1989). *Rickover and the Nuclear Navy: The Discipline of Technology*. Annapolis, MD: Naval Institute Press (pp.3–7).
3   Rickover, H. (19 November 1983). An Assessment of the GPU Nuclear Corporation and Senior Management and its Competence to Operate TMI-1. Retrieved from http://archives.dickinson.edu/document-descriptions/admiral-hyman-george-rickovers-assessment-three-mile-island.
4   Submariners reading this will understand that I did not have a "purple" pen. In submarine slang, to "grape" something off is to get a signature on a qualification card for something you didn't really know or do.
5   The *line of fire* is the path an object, energy, or substance will take once released or discharged; an established line of travel or access; the imaginary straight line from

the muzzle of a weapon in the direction it is pointed, just prior to firing; the path of moving objects or where bullets are being shot or are about to be shot.

6 Organized in 1979, INPO is a private, non-governmental, regulatory body formed by the commercial nuclear industry shortly after the accident at Three Mile Island. INPO develops standards of performance, conducts evaluations of member nuclear utilities against those standards (using a team of specialists), accredits utility training programs, and investigates plant events.

7 The phrase "human performance" was synonymous with "human error." Institute of Nuclear Power Operations (November 1994). *Recommendations for Human Performance Improvements in the U.S. Nuclear Utility Industry.* Atlanta, GA: INPO.

8 The members of the commission included Dr. Aubrey Daniels (independent consultant), Dr. Terrence Lee (University of St. Andrews), Dr. Kim Smart (Bellcore Training and Education Center), Dr. John O'Brien (Electric Power Research Institute), and Dr. Neil Todreas (Massachusetts Institute of Technology), as well as representatives of the Nuclear Regulatory Commission and the nuclear industry.

9 The abbreviation "**Hu**" (the short form for "human performance") was adopted by INPO and the nuclear industry in the mid-1990s. The "HP" abbreviation was already used by the "health physics" personnel in the industry.

10 Institute of Nuclear Power Operations (1997). *Excellence in Human Performance.* Atlanta, GA: INPO. This is an industry publication and is not publicly available.

11 A High-Reliability Organization (HRO) is an organization that operates in an unforgiving social and political environment, rich with the potential for error, where the scale of consequences of error precludes learning through experimentation (avoids learning through trial and error), and where complex processes are used to manage complex technology to avoid failure. Per Dr. Karl Weick, an authority on HROs, high reliability is "the lack of unwanted, unanticipated, and unexplainable variance."

12 See the *Human Performance Improvement Handbook, Volume 1 of 2*, on the U.S. Department of Energy website: https://energy.gov/ehss/downloads/doe-hdbk-1028-2009.

13 Le Coze, J. and Dupre, M. (2008). The Need for "Translators" and for New Models of Safety, in Hollnagel, E., Nemeth, C. and Dekker, S. (Eds.) *Resilience Engineering Perspectives, Volume 1: Remaining Sensitive to the Possibility of Failure.* Boca Raton, FL: CRC Press (Chapter 3).

2 A strategic approach

Primum non nocere

Thomas Sydenham (1624–1689)[1]

Human rather than technical failures now represent the greatest threat to complex and potentially hazardous systems.

James Reason[2]

Human error is likely the greatest source of variation in any human endeavor—an ever-present source of uncertainty that persists throughout your organization.[3] Dr. Reason said that "Wherever there are human beings, there will be human error."[4] When and where people will err cannot be known with absolute certainty—it's not an exact science. This introduces a substantial risk to any organization's operation. The International Organization for Standardization (ISO) describes risk as the "effect of uncertainty on objectives."[5] Therefore, you could conclude that managing the risk associated with the uncertainty of human error is strategic to your organization's long-term safety and prosperity. The word "strategy" (noun) is a plan, method, or series of maneuvers to attain a specific goal or outcome. On the other hand, the word "strategic" (adjective) describes something as essential to success in achieving a specific goal or outcome. I submit that managing the systemic risk of human error in complex, high-hazard operations is essential to preserving not only your people's health and well-being but also the long-term prosperity of your organization, its constituencies, communities, and surroundings.

Events/incidents/accidents happen when assets suffer harm—damage, injury, or loss, usually because assets were not adequately defended. Operational risks arise when people do things—work, using intrinsic hazards to produce products and services. Occasionally, people lose control of hazards through human error. Though human errors happen often, most are trivial, but some can be grievous. It's not important to "prevent" all human error (which is impossible anyway), just the ones that trigger serious negative events.

Events occur when assets suffer harm due to a loss of control of:
- transfers of energy
- movements of mass
- transmissions of information.

Today's high-hazard operations and their related management systems tend to be complex, possessing numerous interdependencies—cause-and-effect relationships—that can easily exceed any one person's capacity to comprehend them all. *Nothing is always as it seems*—there's always something hidden, unknown, incomplete, or otherwise obscure that could lead to trouble, even when in compliance with regulations and "approved" procedures. Consequently, those in direct contact with operational hazards and assets must be able to recognize and adapt (behavior choices) to changing risk conditions to protect assets—a key trait of resilience. The capacity to adapt is necessarily inefficient, but the safety improvements through resilience more than offset the costs in the long run.[6] Please do not misinterpret what I'm suggesting regarding the use of procedures. Technically accurate and usable procedures provide front-line workers the best tool for performing complex operations in high-hazard industries—the stakes are too high to leave the means of producing a work output to individual choice.[7]

Therefore, managers must understand how behavior choices of front-line workers on the shop floor emerge from the designs, functions, values, and norms of their management systems. Understanding and managing the flows and avenues of influence of the organization—the system—on the behavior choices of front-line workers is essential to properly aligning them to minimize the risk of harm and to achieve the business results you want.

> "...if you do not manage human error, human error will manage the organization, always at great cost and often at great danger."
>
> —James Reason
> Video Series, *Managing Human Error*

Human error cannot be left to chance during complex industrial operations. Therefore, I propose an approach that is (1) risk-based, (2) systemic, and (3) adaptive to managing workplace **Hu** during high-hazard operations—to recognize and control the specific **Hu** risks intrinsic in your operations and to verify assets are sufficiently defended should people still lose control. Since H&OP is about managing risk, let's focus on where the risk manifests itself—in the workplace.

Work and the workplace

What is work? From a physics perspective, work is simply the application of a force over a distance—something changes. Work is good when value is created. We enjoy our work when we realize purpose in it, and when we work hard and perform well, we feel the joy of accomplishment and gain a sense of personal worth from it. When exchanged for a wage or salary, work allows an individual to provide for his/her own and help others in need—all good for society. But, occasionally, our work extracts value resulting in damage, loss, or injury.

In the marketplace, industrial operations involve the production of goods and services that are suitable for use or have economic value to a customer base—at a cost that is less than what they would pay to produce it themselves. A variety of processes and methods are used to transform tangible resources (raw materials, semi-finished goods, subassemblies) along with intangible ones (data, information, knowledge, expertise) from their natural, unprocessed state to a finished product state—creating value.[8] Operations associated with production processes require multiple and varied human actions—making, constructing, assembling, manufacturing, inspecting, communicating, operating, or handling material, parts, tools, machinery, and products. Other human activities—not directly involved in the transformation processes—include planning, scheduling, designing, testing, routing, shipping, dispatching, storage, etc. In the marketplace, *work is energy directed by human beings to create value*.

Since physical work involves the use of force during operations, front-line workers require the use of intrinsic (built-in) hazards, most of the time, in various forms of energy. Work—in the battery—occurs when either:

- *energy* is applied to raw materials or intermediate components to create a finished product or provide a service;
- *mass* is transported (whether solid, liquid, or gas) from one place to another;
- *information* is created, processed, stored, transmitted, or communicated to a receiver or recipient.

Notice that these situations are intrinsically associated with normal, everyday work. Transfers, movements, and transmissions occur via various means or pathways initiated by human operators. With today's complex technologies, these pathways tend to be physical in nature—such as the movement of oil through a pipe, heat through an exchanger, flow of electricity through a wire, or delivery of data over electronic and radio networks.

When work occurs—force applied over a distance—something has to change. If work is not performed under control, the change is not what you want; work can inflict harm. Though most hazards are identified before starting work, some appear without warning, which have to be managed in real time. Job hazard analysis, procedures, policies, expectations, training, automation, and facility and equipment designs are ways you use to reduce and control uncertainty and variation in **Hu**. Uncertainty and variation, however, can never be totally eliminated.

If you listed all the hazards you could encounter during your commute to work each day, you would note a few you know are always present (e.g., speed and proximity of other vehicles), but you would also realize that many cannot be anticipated—they are surprises. In the real world, over time, people and things change, knowledge and skills decay, tools and equipment wear out, and assumptions about the environment and local work conditions become increasingly invalid. All along the way, people touch things and with each touch there is the potential to cause harm.

Bottom line: *Work involves the use of force under uncertainty.*

The Hu risk

Generally, safety is the freedom from an unacceptable risk of harm. But, harm to what? Every organization possesses multiple assets, but some are more important than others—people, property, and product as a minimum. Assets and their limitations define the boundaries of harm and what constitutes an event, when those limitations are exceeded. Strategically, it is important to protect your most important assets from harm. Events and accidents are always defined in terms of harm to one or more assets, without which there is no event, except in the case of a serious near-miss (near-hit).

It's crucial for managers of operations to realize an important facet of H&OP relative to the causes of an event. It's not the error that triggers an event that you should be most concerned about. It's the harm to assets that results from error—a loss of control. Harm involves a detrimental change in the state of assets or a serious degradation or termination of the organization's ability to accomplish its mission. Just as human error is unintentional and usually a surprise to the individual, the events that ensue are likewise surprises from an organizational perspective. Managers should not be as concerned with the occurrence of human error as with protecting assets from harm that ensues after the error. Protecting assets from harm motivates the desire to control workplace **Hu**. Ultimately, you want to avoid both (1) losing control and (2) suffering the harm, but each is managed differently.

Let's arrange these elements together in a conceptual model—a way of thinking about risk—that will help you manage the uncertainty of human error in the workplace, to avoid events. Whenever work is performed, as illustrated conceptually in Figure 2.1, three things are present:

- **assets**—things important, of high value, to an organization;
- **hazards**—intrinsic sources of energy, mass, or information used to create value;
- **human beings**—work by fallible people during value-creation processes.

The relationships between **assets**, **hazards**, and **humans** form what I call the "**Hu Risk Concept**." For the remainder of this book, each word used in the **Hu Risk Concept**, including the **Hu Risk Management Model** discussed in a few more pages, is denoted with a bold font as an ongoing reminder of each element's principal role in managing the risk of human error in operations. The **Hu Risk Concept** enhances the recognition of what to manage. The presence of uncertainty due to **human** interactions with **assets** and **hazards** produces risk—risk of harm—an **event**. Because of the potential impact on the organization and the systemic uncertainty that human error poses to your operations, a strategic risk exists. Therefore, managing human error in

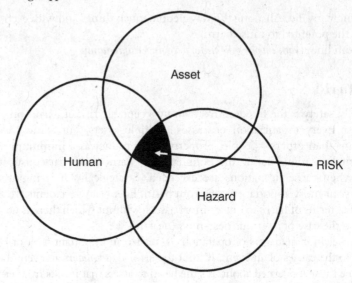

Figure 2.1 The **Hu Risk Concept** illustrates the primary elements demanding attention in managing the risk that human fallibility poses to operations. The interfaces (overlaps) of these elements introduce risk: losing control of an intrinsic **hazard** and harming an **asset**—an **event**

operations becomes as important as managing production—both must be managed together, not separately. Let's look at each element in more detail.

Assets—*what to protect from harm*

Assets include people, product, property, facilities, reputation, equipment—anything of value and important to the organization's reason for being—its mission. Whatever is essential or key to its (1) safety, (2) profitability, (3) reliability, (4) environment, (5) community, and (6) even reputation is of utmost importance to the members of a responsible and honorable organization.

For a business or organization to be sustainable, the **assets** used to create or deliver the company's outputs must be protected. For a commercial airline, the primary **assets** are passengers and its aircraft. Secondary **assets** include schedule reliability and quality of service. For a hospital, **assets** include patients, staff, medications, and its facilities. For a biotech company, it is the drug substance. For a nuclear power plant, it is the reactor core. **Assets** essential to an organization's survival are the rightful focus of H&OP. What happens to **assets** drives the organization.[9]

"If you value it, you will protect it unless you're willing to replace it."
—Dorian Conger
Incident Investigation Expert

In the U.S. Navy Submarine Service, submarines and their various subsystems, including their nuclear reactors, are carefully operated within specific constraints to optimize their safety and reliability, such as depth and speed. Collectively, operators refer to these constraints as the Safe Operating Envelop (SOE). The SOE for any **asset** is the multi-dimensional space of parameters and other conditions under which the **asset** is considered safe. For example, the SOE of a car tire includes wall condition, tread depth, air pressure, temperature, speed, among others. When one or more critical parameters associated with the SOE of an **asset** are exceeded, the likelihood and/or severity of the harm from existing **hazards** increases.[10] Critical parameters are the vital signs of the health or illness of equipment and processes. A fragile **asset** is one that can be easily harmed by handling, jostling, stress, contact, collision, abruptness, neglect, etc. Using more robust **assets** reduces the potential for harm, but may be more expensive. **Assets** are generally protected from harm during operations by ensuring certain critical parameters remain within the confines of their SOE (design bases) using procedures and built-in defenses.

Anything that can harm or damage key **assets** must be taken seriously by the organization's managers at every level—as part of its everyday core business processes. Examples of undesirable business outcomes include various forms of the following:

- products or services that harm customers;
- people killed, disabled, injured, or infected with a disease;
- product or property lost, defective, damaged, or destroyed
- environment contaminated, spoiled, or ruined;
- reputation defamed or tarnished;
- information lost, corrupted, or stolen (intellectual property or trade secrets);
- value lost due to excess waste or scrap and late deliveries.

Obviously, a company that continues to experience any of these problems—**events**—for extended periods will not stay in business very long.

Hazards—*built-in sources of harm*

Various types of **hazards** are intrinsic to industrial operations—built-in sources of energy, mass, and information that are associated with a particular domain of operations. They are necessary to accomplish the organization's work. Intrinsic physical **hazards** used in industrial operations make harm a real possibility. Generally, **hazards** are intrinsic either to the material or to its conditions of use. For example, in the maritime industry, intrinsic **hazards** can include water, deep water, rocks and shoals, corrosion, currents, tides, and the presence of other vessels. Deep-draft merchant vessels cannot operate safely in shallow water near shoals. In aviation, they include gravity, elevation, weather, terrain, and other aircraft. The following list of energy forms (with examples in parentheses) shows some of the more common intrinsic **hazards** used in operations:[11]

- electricity (power sources (potential), electrostatic charges);
- kinetic (rotating machinery, flywheels, moving equipment, flow, velocity);
- chemical (acids, corrosion, fire, dust, lead);
- gravity (elevated work, bulk storage at heights, hoisting and rigging);
- thermal energy (steam, fire, hot surfaces, bright lights);
- compressed fluids (high pressure gases, hydraulics, vertical columns);
- toxic or inert gases (phosgene, carbon monoxide, confined space);
- explosives (hydrogen, natural gas, gasoline fumes);
- acoustic (noise and vibration);
- radiation (x-rays, ionizing, lasers);
- biological (viruses, bacteria, fungi, animals, insects);
- information (defective software updates, out-of-date prints, unsecured networks, missing or inaccurate procedures, corrupt data, lax security protocols);
- people (fallible decisions, imprecise actions and movements).

Safety and control of hazards depends on the recognition of hazardous conditions, so as to prevent unexpected energization, startup of equipment and machinery, or the release of stored energy that could injure workers. This means people should possess an in-depth technical knowledge of the technology they work with, and be able to recognize various hazardous energy sources, their types, and magnitudes present.

"If it has the capacity to do work, it has the capacity to do harm."
—Dorian Conger
Incident Investigation Expert

We tend to assume **hazards** are stable—that they are always present and knowable in advance. Most are, some aren't. Occasionally, unanticipated **hazards** encountered during work appear gradually or unexpectedly, not unlike the normal rise and fall of sea levels or from storm surges spawned by earthquakes and distant storms. To sustain the safety of your **assets** over the long term, front-line workers have to be capable of managing the surprise **hazards** when they occur—able to adapt.

Events—*harm to* assets

Events have many names, such as incident, accident, deviation, nonconformance, etc. In other contexts, the term "**event**" can refer to everyday occurrences such as a birthday party, a wedding, a concert, or a baseball game. Here, I refer to the negative meaning of the word **event**—an undesirable occurrence involving harm (injury, damage, or loss) to one or more **assets** due to uncontrolled (1) transfers of energy, (2) movements of mass, or (3) transmissions of information, as noted earlier. **Events** are not entirely unpredictable—unpredictability resides in their timing and location, not so much their causes.[12] I believe that causes of **events** that have not happened can be found and eliminated.

"Day after day, year after year, nothing much goes wrong, lulling managers and workers into a sense of security and a belief that what they do day after day is safe. Events are always surprises."

—Derek Viner
Author: *Occupational Risk Control* (2015)

The anchor point in any **event** occurs at a point in time when control is lost over the damaging properties of energy, mass, and information—when the destructive potential of intrinsic **hazards** is unleashed uncontrollably because of a loss of control or the absence of adequate protection.[13] In every **event**, defenses had to fail in some way or were circumvented. See Figure 2.2. An **event's** severity is not so much a result of the worker's error, it is more a function of the amount of energy absorbed by an **asset**. An **event's** severity is measured in terms of the degree of injury, loss, or damage to one or more **assets**—a function of the magnitude, intensity, or duration of the **hazard's** damaging release. For example, an automobile accident triggered by the same driver's error in either case is more severe at high speeds than at low speeds. It should be apparent that the robustness of defenses—barriers and safeguards—determines how bad, or how benign, outcomes are after an error.

The harm that ensues after a loss of control, is more the result of ineffective defenses, where the barriers or safeguards were either missing, ineffective, or by-passed. As illustrated in Figure 2.1, the **Hu Risk Concept**, sustaining safety in operations is essentially a control problem—control of (1) human variability during high-risk activities, and (2) the intrinsic hazardous processes used during work.[14]

Human fallibility—potential for losing control

The occurrence of an **event**—the onset of injury to, damage to, or loss of one or more **assets**—strongly indicates that control over the damaging properties of

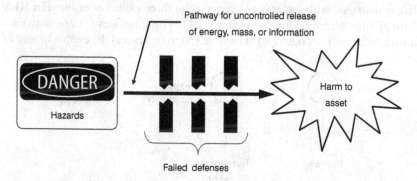

Figure 2.2 An **event** occurs due to an "uncontrolled" transfer of energy, movement of mass, or transmission of information—failures in defenses—that allows harm to occur to an **asset**. (Adapted from Figure 1.1 in Reason, J. (1997). *Managing the Risks of Organizational Accidents*. Aldershot: Ashgate.)

intrinsically hazardous processes was lost—usually triggered by some human failing at the controls.[15] Variations in behavior, due to human error, lead to variations in results. The notion of human error is explored more deeply in Chapter 3.

Yes, people are **hazards**, but they are also heroes. Creativity and fallibility are two sides of the performance coin. Through creativity, people possess the ability to adapt to unexpected risky situations. Because of this innate human characteristic, people are able to create safety for **assets** in situations not previously anticipated by either managers or designers. Creativity springs from technical expertise. Expertise and the ability to adapt are important features of RISK-BASED THINKING, which will be addressed more fully in Chapter 4.

In the healthcare industry, the overarching, guiding principle is to "first, do no harm." That's exactly the mindset that managers and leaders need in high-hazard operations. In the following paragraphs, I describe a management model—derived from the **Hu Risk Concept** that helps you and your organization "do no harm." A model-based approach to managing risk provides managers with the means to be proactive in avoiding harm, by helping them understand how their systems influence performance and its defenses. An event-based approach—learning late by reactive reporting and **event** analysis—will only get you so far.

Hu risk management model

The battery of **Hu**—the confluence of **assets**, **hazards**, and **Hu** illustrated conceptually in Figure 2.1 occurs during work. Using the **Hu Risk Concept** as a springboard, a more practical form, depicted in Figure 2.3, suggests more specifically what to manage—**pathways** and **touchpoints**. It is a simple, highly transferable model that helps pinpoint the work-specific interfaces (or combinations) to pay attention to in order to avoid an **event**.

Pathways and touchpoints

Risk is managed at the interfaces between the three elements of our **Hu Risk Concept** (overlapping circles in Figure 2.1). The interfaces (interactions) are illustrated by the plus signs (+). The first + sign represents the establishment of

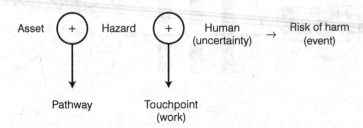

Figure 2.3 The **Hu Risk Management Model** pinpoints the two work-related interfaces—**pathways** and **touchpoints** (denoted by the plus (+) signs), that must be controlled to minimize the risk of harm during operations

a **pathway** for either the transfer of energy, movement of mass, or transmission of information between an operational **hazard** and an **asset**, where only one human action (or equipment failure) is needed to either create value (under control) or cause harm (out of control). A **pathway** for harm exists when a **hazard** is poised in such a way as to expose an **asset** to the potential for a change in state—a vulnerability for good or for bad. Remember, the original intent of these interfaces is to add value through work. However, whenever there is an opportunity to add value, there is an associated risk to do harm—to extract value. Derek Viner, in his book, *Occupational Risk Control*, refers to such exposures as "vulnerability **pathways**."[16] For example, a firearm that has a bullet loaded in the chamber, with the safety off and the hammer cocked is poised to discharge. A person would be exposed to death or severe injury if standing in the *line of fire* in front of the muzzle of a firearm in such a condition, especially if the firearm is wielded by another person with a finger on the trigger (a **touchpoint**). Other examples of potential **pathways** include the following situations:

- a toddler standing near the edge of the deep end of a swimming pool;
- the presence of fuel, oxygen, and a nearby ignition source, such as a match;
- an electrical potential (voltage) controlled by a single switch or circuit breaker;
- a person standing on the curb of a busy street;
- a vehicle parked on a steep hill secured only by the automatic transmission in Park (P) and with the front wheels aimed straight ahead;
- accessibility to proprietary information controlled by a simple password by the wrong persons;
- a digital control system (DCS) aligned to start a series of automated process sequences by an operator with a finger on mouse prepared to click "Go" or to depress the Enter key;
- a phishing e-mail message open on the screen of your personal computer with the cursor hovering over a link to a malicious website.

Recall the photo of the ironworkers in Figure 0.1. It vividly illustrates the elements of the **Hu Risk Management Model**. This photograph highlights the three distinct elements of risk: **asset**, **hazard**, and **human**. Obviously, the **assets** are the people, the ironworkers. What makes the photograph stunning is the obvious **hazard**—the beam the workers are sitting on is more than 800 feet about the streets below. The apparent absence of barriers (nets or body harnesses) to prevent them from falling makes it doubly fearsome, which is accentuated by the fact that people are fallible, including the photographer. **Pathways** are particularly important because the potential for harm is now dependent upon either a single human action or equipment malfunction (failure). Front-line workers must be wary of the creation and existence of **pathways**.

The second + sign represents a **touchpoint**. A **touchpoint** involves a human interaction with an object (whether an **asset** or a **hazard**) that changes the state of that object through work. **Touchpoints** involve a force applied to an object over a distance, using tools or controls of hazardous processes. After performing a **touchpoint**, "things are different," and usually things can't be put back the way they were. The control of a **touchpoint** is most important when a **pathway** for energy, mass, or information exists between an **asset** and a **hazard**. A **touchpoint** includes all the following characteristics:

- *human action*—bodily movements, exerting a force;
- *interaction with an object*—physical handling—force applied to an object;
- *work*—force applied over distance;
- *change in state*—changes in parameters that define the state of the object.

Human beings possess the capacity to direct energy, to move things with their hands, feet, and body. Because of our innate human fallibility, **touchpoints** bring about uncertainty related to the actions performed—whether under control or out of control, and their outputs, whether for good or for bad. Quality Assurance people call this variation. In the photograph of the ironworkers in Figure 0.1, the **assets**, **hazard**, and **pathway** are clear. However, the all-important **touchpoint** may not be so apparent. They're sitting on them—their backsides. If the workers lean too far forward or backward, either way they could lose their balance and fall to their deaths.

Both the creation of **pathways** and the occurrence of **touchpoints** (work) are normal and necessary activities of everyday operations. The occurrence of **touchpoints** after creating a **pathway** tend to be critical to safety. If a **touchpoint** is performed in error, the performer loses control and harm—an **event**—is likely to occur. This suggests that defenses must be built into the facility design, production processes, procedures, and expectations to protect **assets** from errant operations, while **pathways** exist. Otherwise, harm is likely.

The risk of an **event** is managed through the prudent deployment of controls, barriers, and safeguards to (1) lessen the chance of human error at important **pathways** and **touchpoints**, and (2) protect **assets** from harm. This assumes no modifications or alternatives exist to the **assets** and **hazards** used during operations. As mentioned earlier in the Introduction, the scope of this book is limited to managing H&OP with what you have.

The building blocks of H&OP

Strategically, what should managers "control?" As stated earlier, H&OP is all about managing the risk human error poses to an organization's operations. In light of the **Hu Risk Management Model**, the building blocks of H&OP suggest specifically what managers should pay attention to and what to do. As illustrated in Figure 2.4, H&OP involves three core operational functions and three management support functions. The core operational functions that involve daily risk management include:

1 RISK-BASED THINKING—adapting to real-time risks in the workplace.
2 CRITICAL STEPS—ensuring the right things go right at critical phases of work.
3 SYSTEMS LEARNING—detecting and correcting ineffective defenses and related system weaknesses.

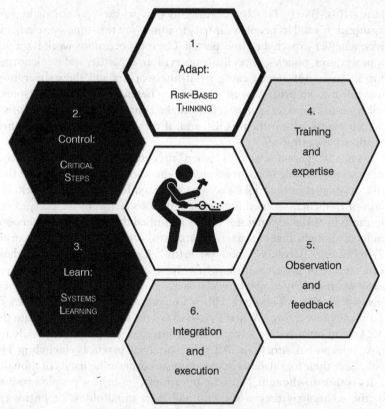

Figure 2.4 The Building Blocks of Managing H&OP. Adapt: Risk-Based Thinking; Control: Critical Steps; and Learn: Systems Learning serve as the core risk-management functions of H&OP (black hexagons). However, without direct line manager engagement, H&OP will not work: Training and expertise, Observation and feedback, and Integration and execution (gray hexagons)

Cells 4, 5, and 6 are important management practices that support the effectiveness of cells 1 through 3. These include:

4   *Training and expertise*—ensuring front-line personnel possess technical knowledge and skill to exercise Risk-Based Thinking and to recognize and control Critical Steps.
5   *Observation and feedback*—managers spending time on the shop floor, creating learning opportunities for both workers and themselves.
6   *Integration and execution*—enabling Risk-Based Thinking as a way of doing work, and promoting managerial accountability for the follow-through of all H&OP functions.

Collectively, the six functions form the building blocks of H&OP, which are described individually in more detail in the following paragraphs and chapters.

1   *Adapt*—RISK-BASED THINKING. This function focuses on enhancing the organization's and its people's capacity to adjust—in real-time—to changing risks, whether expected or unexpected. On most occasions work is guided by procedures, policies, checklists, supervision, expertise, and work norms. But, in some unforeseen cases, front-line workers and their supervisors may run into surprising work situations. Therefore, front-line personnel will have to "think on their feet" to make adjustments to such surprises to keep **assets** within their SOEs, and, if they can, still accomplish their organization's purposes.

It's not "if" people adapt; it's "when" they adapt—they must adapt. And, most people are able to adapt because of the knowledge and expertise they have developed in their jobs over several years. RISK-BASED THINKING is a form of resilience—something a system does to verify/ensure safety exists. Research in resilience engineering has identified the following cornerstone habits of thought that characterize successful organizations: (1) anticipate, (2) monitor, (3) respond, and (4) learn.[17] Collectively, I refer to these cornerstones or patterns of thought as RISK-BASED THINKING. *Although you cannot anticipate every possible harmful scenario, you can be ready for almost anything through RISK-BASED THINKING.* This is a bold assertion, but it's the strength of a resilient workforce. When RISK-BASED THINKING is integrated into the DNA of an organization's way of doing business, people will be ready for most unexpected situations. All safety-oriented practices, including **Hu** tools, have their foundations in one or more cornerstone habits of thought.

Its cousin-in-thought, "chronic uneasiness," enhances people's respect for the technology they work with and their mindfulness of **pathways** between **assets** and **hazards** and impending transfers of energy, movements of mass, and transmissions of information. Top performers think before acting to prove to themselves that an **asset's** safety exists before doing work. Chapter 4 explores this building block in greater depth.

2   *Control*—CRITICAL STEPS. This function focuses people on ensuring the right things indeed go right the first time, every time. A CRITICAL STEP is a human action that will trigger immediate, irreversible, intolerable harm to an **asset**, if that action or a preceding action is done improperly. The more work, the more **touchpoints** in a given time means more errors occur for the same period—more shots on goal. It is likely more **events** will occur. The more errors, the more often **events** occur. As you can see, the tempo of work activities tends to drive the frequency of events; the number of people required, how often people do work, the amount of time involved, and the number of human actions.

Events are triggered at CRITICAL STEPS. Identifying and controlling **Hu** at CRITICAL STEPS, minimizes unwanted variation in **Hu**—avoiding a loss of control. Specifically, managers would do well to help workers and supervisors become keenly aware of CRITICAL STEPS and related error traps that aggravate the potential for error during those actions. Workers and supervisors know what to pay close attention to and what to do to exercise positive control of these critical **touchpoints** that trigger transfers of energy, movements of mass, and transmissions of information during operations. Chapter 5 explores this building block in great detail.

3   *Learn*—Systems Learning. The safety of any system depends on the presence, integrity, and robustness of its defenses. Systems Learning involves the detection and correction of weaknesses in an organization's system that diminish the integrity, presence, or robustness of defenses, weakening the control of **Hu**, and/or inhibiting the protection of **assets**.

Error prevention has been the traditional focus of managers wanting to "improve **Hu**." However, I hope that you are realizing that the emphasis on error prevention is overblown. Only defenses (barriers and safeguards) built into systems, processes, structures, and components minimize the severity of **events**.[18] The severity of **events** is not a result of the trigger mechanism—human error. It's a result of the marshalling of energies, an **asset's** susceptibility to harm, and the integrity of defenses built into the system. Finding and correcting faulty defenses and eliminating hidden **hazards** in the workplace tend to minimize the severity of **events**. Not doing so leaves the system vulnerable and, therefore, unsafe—even though **events** may not be occurring.[19] Systems Learning is effective only when line managers are accountable for the ongoing discovery and elimination of system weaknesses that inhibit the effectiveness of defenses.[20] A deep dive on Systems Learning happens in Chapters 6 and 7.

4   *Training and expertise.* Training, experience, and proficiency builds expertise— intimate understanding and skill associated with a technology, combined with a *deep-rooted respect* for the dangers of that technology. Training also informs people of their capabilities and limitations as human beings as well as a means to moderate their fallibility. Risk-Based Thinking and chronic uneasiness are integrated into technical training programs. Expertise is the bedrock of Risk-Based Thinking. Knowledge and skill is perishable—it has a half-life. Therefore, training must be ongoing—not a one-and-done activity. Training is discussed in Chapter 8.

5   *Observation and feedback.* Because of its importance and usefulness to line managers as a Systems Learning tool, observation and feedback is called out separately. Observation provides opportunities for two-way feedback via face-to-face interactions between line managers and front-line personnel. Through observation and feedback, managers see firsthand what is happening in real time and what front-line workers have to work with—what they inherit from the system, including unworkable and ineffective procedures, tools, and expectations. Managers see with their own eyes what workers do to accomplish work, whether according to expectations or otherwise. Workers and supervisors receive feedback about their performance in the workplace, and managers receive feedback about their systems. Chapters 7 and 8 discuss the importance and conduct of observation and feedback.

6   *Integration and execution.* Eventually, Risk-Based Thinking, Critical Steps, and Systems Learning are incorporated into all facets of operations and the

organization. It becomes part of the daily core business—it is not optional. Safety and production go hand-in-hand—they're not separate activities. Safety occurs while you work. Integration enables RISK-BASED THINKING in various organizational functions as well as operational tasks and work processes—it becomes a way of thinking and doing work. Execution is a systematic and disciplined management process of getting things done.[21] Managers can't simply hope front-line personnel will use **Hu** tools (see Appendix 3) to avoid errors—the risks must be properly managed. Implementing H&OP and RISK-BASED THINKING requires management, leadership, commitment, and accountability. Chapter 9 is devoted to this building block.

The primary benefit of systemwide deployment of H&OP is sustained periods of success. A declining trend in the frequency and severity of **Hu events**, the reduction of costs and risks to the organization, and the ever-increasing adoption of safe practices, among others, are manifestations of H&OP. Applying RISK-BASED THINKING and reducing the occurrence of errors at CRITICAL STEPS tends to drive down the frequency of **events**. The severity of the **events** that still occur are minimized by ensuring barriers and safeguards are in place and effective in protecting the organization's **assets**.

I believe that institutionalizing H&OP will not only save lives through improved safety and reliability, but it will also save livelihoods by improving the organization's productivity (production/unit time) in the near term, and profitability (production/unit cost) over the long run.

Things you can do tomorrow

1   Discuss with colleagues or your management team what is meant by "safety." Discuss how safety is managed. Ask your colleagues or management how they know safety exists if no **events** are occurring.
2   Do managers believe that **events** are unanticipated, sudden, surprising, unpredictable, and "caused" by "unsafe" worker behavior? Such beliefs foster resistance to the idea that **events** can be prevented.
3   During any production-focused meetings, observe whether safety is separate or part of the conversation. Do meetings start with a "safety moment," followed by the "real work?" Or, is protection of assets part and parcel with talk about the production objectives?
4   Using a frequent high-risk operation as an example, discuss with the management team how the risk of human error could be better managed using the **Hu Risk Management Model**.
5   Identify operations or work activities on the current schedule that are high-risk or potentially costly if control is lost. Consider means to avoid losing control. Identify contingencies, "STOP-work" criteria, and ways to protect assets if control is lost.

Notes

1 "First, do no harm." The medical maxim is commonly and mistakenly thought of as the Hippocratic Oath. This specific maxim and its Latin phrase was attributed to the English physician Thomas Sydenham (1624–1689) in a book by Thomas Inman. See Inman, T. (1860). Hays, I. (ed.) "Book review of *Foundation for a New Theory and Practice of Medicine.*" *American Journal of the Medical Sciences*, Philadelphia, PA: Blanchard and Lea (pp.450–458).

2 Reason, J. (1995). Understanding Adverse Events: Human Factors. *Quality in Health Care,* 4.2:80-89. DOI: 10.1136/qshc.4.2.80.

3 Flin, R., O'Connor, P. and Crichton, M. (2008). *Safety at the Sharp End: A Guide to Non-Technical Skills.* Farnham: Ashgate (p.1).

4 A quote associated with Dr. James Reason in his video series, "Managing Human Error."

5 International Organization for Standardization (2009). ISO 31000. "Risk Management."

6 Van Dyck, C., Frese, M., Baer, M. and Sonnentag, S. (2005). Organizational Error Management Culture and Its Impact on Performance: A Two-Study Replication. *Journal of Applied Psychology.* Vol. 90. No. 6. (pp.1228–1240). DOI: 10.1037/0021-9010.90.6.

7 Marx, D. (2009). *Whack-a-Mole: The Price We Pay for Expecting Perfection.* Plano, TX: By Your Side Studios (p. 67).

8 Retrieved from http://www.businessdictionary.com/definition/production.html#ixzz3q3cbPuXq

9 Reason, J. (1997). *Managing the Risks of Organizational Accidents.* Aldershot: Ashgate (p.3).

10 Corcoran, W.R. (August 2016). An Inescapable of the Safe Operating Envelope (SOE). *The Firebird Forum.* Vol. 19, No. 8.

11 Viner, D. (2015). *Occupational Risk Control: Predicting and Preventing the Unwanted.* Farnham: Gower (pp.70–72).

12 Ibid. (p.27).

13 Ibid. (p.46).

14 Leveson, N. (2011). *Engineering a Safer World.* Cambridge, MA: MIT (pp.67, 75).

15 Viner, D. (2015). *Occupational Risk Control: Predicting and Preventing the Unwanted.* Farnham: Gower (p.112).

16 Ibid. (p.43).

17 Hollnagel, E. and Woods, D. (2006). Epilogue: Resilience Engineering Precepts. In Hollnagel, E., Woods, D. and Leveson, N. (eds.). *Resilience Engineering: Concepts and Precepts.* Farnham: Ashgate (pp.349–350) and (2009) *Resilience Engineering Perspectives, Vol. 2.* Farnham: Ashgate (pp.117–133).

18 Idaho National Engineering and Environmental Laboratory (March 2002). "Review of Findings for Human Contribution to Risk in Operating Events," (NUREG/CR-6753). Washington, DC: U.S. Nuclear Regulatory Commission.

19 Retrieved from www.fra.dot.gov/downloads/safety/ANewApproachforManagingRR Safety.pdf

20 Reason, J. and Hobbs, A. (2003). *Managing Maintenance Error, A Practical Guide.* Farnham: Ashgate (pp.91, 96).

21 Bossidy, L. and Charan, R. (2002). *Execution: The Discipline of Getting Things Done.* New York: Crown (pp.21–30).

3 Fundamentals of human and organizational performance

Saying an accident is due to human failing is about as helpful as saying that a fall is due to gravity. It is true but it does not lead to constructive action.

Trevor Kletz[1]

Individuals working in organizations make errors every day and every hour and (sometimes) make multiple errors in the span of a minute.

David Hofmann and Michael Frese[2]

You don't have to live very long to know that people aren't perfect. People have various talents and capabilities, and some work harder than others. Everyone has free will to make choices, but at the same time everyone is inescapably fallible.[3] People are known to hurt themselves and others, to break things, and to lose things—among a myriad of other mistakes and blunders. Society for the most part is ashamed of error. It is often the spotlight of public ridicule and shame. Often, it is considered the opposite of good performance. Although most people would agree that "to err is human," at the same time, they subconsciously believe that human error is a choice—motivated. Many people, especially in western societies, think there is a flaw in a person's character when he or she errs, especially those triggering serious harm. It's convenient to assume that people, who act "carelessly" or "let something slip," must lack the proper motivation to do their jobs correctly.

Human performance is *not* simple. Performance always involves complex relationships: relationships between people and processes, people and technology, people and machines, and people and other people.[4]

As human beings, we all are **hazards** to our own endeavors, and you are responsible for managing this risk. However, you cannot manage what you do not understand.[5]

Without a fundamental knowledge of human performance, you—the line manager or first-line supervisor—will struggle to achieve any consistency in human and organizational performance, and progress in real safety over the long term will be frustrating. This is not a training problem but an educational challenge for managers—their understanding. To be consistent and effective in managing human and organizational performance and its associated operational risks, it's important to understand some of the fundamental science around human performance, especially in industrial organizations. This chapter will help you better understand the fundamentals and the particular significance of human error in high-hazard operations—the prerequisite knowledge for managing the human error risk in an organizational context.

The notion of human error

Society at large conflates error with sin. *Error is not a choice and is always unintentional.* Error tends to break things—sin, on the other hand, tends to break relationships and erode trust. Sin for the most part is intentional, mostly born out of personal pride and selfishness. Everyone, especially line managers, must realize that there is no ill will associated with human error—*it's not a moral issue!*[6]

Many people—including managers in high-hazard industries—believe workers should always be able to judge between good and bad, right and wrong, correct and incorrect. But this perspective is a flawed—and dangerous— assumption within the context of a complex, technology-intensive operation. Good baseball players fail to reach base around 60 to 70 percent of the times they enter the batter's box, and all good baseball teams work around that fact. Fortunately, work in the commercial world is not so error prone, where work most often adds value instead of extracting value. Most often, the reliability of **Hu** is nominally in the vicinity of 99 percent (or better), depending on local conditions.[7]

Consequently, there is a tendency to misunderstand human error at work, especially when serious consequences ensue. Frequently, industry line managers count errors as a measure of bad or unprofessional performance, and they set goals of "zero errors." It's common to not realize the occurrence of an earlier error until after suffering some damaging consequences. Other times we recognize mistakes without sustaining any unwanted outcomes, and we exclaim to ourselves, "Whew! That was a close one!" (near hit or close call).

In this book, I define human error as a behavior (human action) that unintentionally deviates from a preferred behavior for a given situation—a *loss of control.*[8] It's important to accept the premise that when a person's action unintentionally deviates from a preferred action, there is no intent to cause harm. Generally, people come to work wanting to do a good job, to be productive and effective. Unless there is clear evidence to the contrary, we must presume people are honorable and trustworthy in their work.

The "preferred" behavior is defined by the work situation in which **assets** are exposed to hazardous processes in the workplace. Often, the preferred behaviors are specified by a procedure, an operator aid, or a checklist in your specific expectations for standard, repetitive work. However, sometimes the preferred behavior is not so apparent, especially when the work situation does not clearly match the one presumed by the system's designer or the procedure's author. In such circumstances, the system's components may have responded or interacted in ways not previously anticipated by the designer or author. Occasionally, the front-line worker must make a choice to either protect key **assets** from harm or accomplish the job's production objectives. Which way will the individual lean? More on this conundrum later when I address "conservative decision-making" in Chapter 4.

"...human fallibility is like gravity, weather, and terrain, just another foreseeable hazard. Error is pervasive... What is not pervasive are well-developed skills to detect and contain these errors at their early stages."

—Karl Weick and Kathleen Sutcliffe
Co-authors: *Managing the Unexpected* (2015)

Some people might feel, maybe strongly, that the term "human error" should be dropped from our vocabulary and should not be a focus of management scrutiny. In the context of daily high-risk business operations, I genuinely believe this is unwise. I deeply believe and know from personal experience that people do indeed err—sometimes triggering injury, damage, loss, or harm (recall my personal close call in Chapter 1). Human error in specific high-risk situations needs to be controlled. With that said, let me say with equal conviction that I do not think of front-line workers as a problem to be fixed, eliminated, or improved. If you have studied human performance, high reliability, or resilience engineering, you will no doubt have come across the phrase "human performance improvement." This book is *not* about improving anyone. H&OP is all about managing the risk to **assets** associated with human error during operations—a reality that line managers must be ready and able to respond to daily and systematically.

It is impossible to conceive a scenario where an entire organization is completely free of human error. Human error is an intrinsic characteristic of any system that employs people—a residual risk that cannot be totally eliminated. Even if workers are eliminated by automated processes, human beings are still involved in the process' design, installation, and maintenance. People are fallible, and as human beings, they will lose control somewhere, sometime. Managing human error is about managing risk—risk to **assets**. Managing risk requires a systems perspective, understanding how people's choices emerge from the system within which people act (the work context), and how an **asset's** defenses fail to provide the protection when needed.

Basics about behavior, error, and violations

Work requires people to touch things—to handle, manipulate, or alter things. Work comprises a series of human decisions and actions that eventually change the state of material or information to create outputs—**assets** that have worth in the marketplace. Work is behavior and is the key to success for any organization. Without the behavior of people—their actions—nothing is accomplished in business. Because of its importance to managers and their success organizationally, I devote an entire chapter to the topic of Managing Human Performance in Chapter 8.

Human error is also behavior. Error is an action that unintentionally deviates from a preferred behavior, such as misspelling a word as you type—whether an act of commission or omission. Error is normal—a fact of life, a natural part of being human. Human fallibility is an intrinsic characteristic of the human condition—to err is human. The ways we, as human beings, err is nothing new—slips, lapses, trips, and fumbles—unintended actions—usually take on one of the following forms:

- timing—too soon / too late
- duration—too long / too short
- sequence—out of order
- selection—wrong object
- force—too much / too little
- direction—wrong direction
- speed—too fast / too slow
- distance—too far / too short.

Mistakes, on the other hand, are planning errors that don't match reality and lead to unwanted results—intended actions for the wrong reasons. They involve mismatches between prior intentions (before execution) and intended work outputs or results. Though mistakes are choices, intentional in nature, the outcomes are unintentional. Violations, in contrast to error and mistakes, involve deliberate, intentional deviations from a known rule, law, procedure, or expectation. Note that the procedure or expectation may be wrong for a unique situation not anticipated by the author—adhering to the procedure would cause harm. In such situations, the so-called violation may be what you want the person to do—adapting to protect an **asset** from harm. Error, mistakes, and violations are all behaviors.

Trying to "cure" people of human error is like trying to stop an infant from crying—it's not going to happen. You must learn to live (to operate) with it. Error is not a choice a person consciously makes. In an operational environment, I think it is more constructive for managers to think of human error as a loss of control. Error tends to reduce the margin of safety, and increase the likelihood of **events**. But, under the "right" conditions, an error could result in harm, injury, damage, or loss, if you—and your system—are unprepared for the moment-by-moment reality of human error.

Because of the limitations of human nature, human fallibility is a chronic source of uncertainty and variation during operations. Depending on which books, reports, and articles you study, error rates for human beings range anywhere from 1 to 15 errors per hour. A study by the Civil Aviation Safety Authority of Australia suggests people make roughly 50 errors a day (three to four errors per hour).[9]

Dr. Trevor Kletz in his book, *An Engineer's View of Human Error*, approximates the "nominal error of commission" to around 3×10^{-3} or 99.7 percent reliability; roughly 1 error in 1,000 attempts, when engaged in a deliberate activity.[10]

With such probabilities, human error is likely the greatest source of variation in your operation—an enduring, residual risk. The uncertainty of when and where human error occurs cannot be ignored and must be managed. However, we cannot be so concerned with avoiding errors as about avoiding harm to **assets**. Do not be overly concerned with preventing all errors—as mentioned before, it's truly an impossibility. People are making mistakes all the time, even as you read this sentence. This can be disheartening to an operations manager. But, let me ask you a simple question. Can you pinpoint specific operations (at **pathways** and **touchpoints**) that possess the capacity to cause injury, damage, or loss? Yes, I know you can. And that provides the opportunity to manage them proactively. Strategically, you want to manage the risk—the uncertainty.

But how do you manage something that is so fickle and resistant to control as human error? It is helpful to understand the results of error. Dr. James Reason said, "It is important to distinguish between two kinds of error: *active errors*, whose effects are realized almost immediately, and *latent errors*, whose adverse consequences lie dormant within the system for a long time only becoming evident when they combine with other factors to breach the system's defenses."[11]

Active errors and latent errors are managed differently.

Active errors

Active errors have an immediate, direct impact on safety (**assets**). The company (and maybe the persons involved) is worse off after an active error than it was before. Active errors are associated with the loss of control of operational **hazards** that can cause immediate and serious consequences, and these are the errors you should want to anticipate and avoid in the workplace. Since all people make mistakes, you want to catch active errors before they cause serious harm.[12]

Instead of adding value, active errors extract value—exacting an immediate cost to the organization. You can anticipate and avoid active errors by pinpointing the creation of **pathways** between important **assets** and their intrinsic **hazards**. Then, control the when, where, and how people interact with those **assets** (**touchpoints**) and the related defenses (of **assets** and of intrinsic **hazards**).

Active errors, which have immediate negative consequences, tend to drive the frequency of events. The more active errors, the more events that occur. The following examples illustrate this effect:

- closing a circuit breaker onto a grounded system, blowing up a transformer;
- entering a tank from the top entrance with an oxygen deficient atmosphere;
- opening the incorrect circuit breaker, de-energizing vital equipment;
- rupturing a pressurized natural gas line while digging a trench;
- swallowing a bite from a cheeseburger when one is allergic to cheese;
- giving a frail infant 10 times the recommended dosage of a medication.

Bottom line: *Active errors trigger events*.

Latent errors

Unlike active errors, nothing happens, immediately, after a latent error. A latent error is an action, inaction, or decision that creates an unwanted (yet hidden) condition that goes unnoticed at the time it occurs—causing no immediate or apparent harm. The word latent means hidden, dormant, or covert. The difficulty with latent errors is that it is nearly impossible to know when they occur, since there is no immediate feedback to alert the performer. Latent errors are silent behaviors (that occur a point in time), but they leave evidence in their wake—latent conditions— organizational weaknesses in management systems, inadequate training, poor work plans, poor human–computer interface design, unsafe values and priorities, etc. Additional weaknesses include at-risk conditions in the workplace such as poor workplace layout, inaccurate or incomplete procedures, unworkable or missing tools, lack of supervision of high-risk activities, and clumsy automation.

> *Note.* James Reason remarked in his book, *Human Error* that "…it is neither efficient nor effective to prevent 'latent errors', but to quickly find and detect the unsafe outcomes of these errors."[13] A relentless posture toward SYSTEMS LEARNING offers a sound approach for detecting and correcting latent conditions (see Chapter 7).

While errors by front-line workers tend to be obvious, errors by knowledge workers and managers tend to be difficult to detect and can lead to conditions that have cascading effects later—such as reductions in staffing, postponing or canceling training, or extending maintenance periodicities for equipment. Front-line workers also make latent errors, such as misconfiguring equipment, incorrectly recording data, or using wrong materials in maintenance. Also, it should not be surprising that latent errors usually occur when people make rushed operational decisions with incomplete information—not an unusual occurrence in the marketplace. Latent errors can be committed by anyone, as illustrated by the following examples:

- Engineers perform key calculations incorrectly that slip past reviews and end up invalidating the design basis for safety-related equipment.
- Maintenance personnel incorrectly install a sealing mechanism that is not discovered until the equipment fails to function during later use.
- Technical staff (such as engineers and procedure writers) embed inaccuracies in procedures, policies, drawings, and design documentation.
- Operators unknowingly alter the configuration of physical plant equipment, such as mispositioning a valve, placing a danger tag on the wrong component, or neglecting to affix a deficiency label on an improperly calibrated instrument.

As you can see, latent conditions arise from the errors, misjudgments, and miscalculations made by anyone.[14] Latent conditions may exist for mere moments or persist for years, even decades, before they combine with either active errors or other local circumstances to trigger an **event**. Once a latent condition is created, there is a risk that the condition will combine with future errors or other conditions that could trigger an **event**. The time-at-risk clock has started. As a manager, your objective is not so much to prevent latent errors—but to find and correct the latent conditions left behind, minimizing the vulnerability—the time-at-risk. This is why SYSTEMS LEARNING is so important (see Chapter 7).

Latent errors create unsafe or at-risk conditions that make the workplace and the physical plant vulnerable to errors during future operations. In their book, *Errors in Organizations*, Rangaraj Ramanujam and Paul Goodman accentuate the dangers of latent errors.[15]

...latent errors [and the adverse latent conditions they create] can potentially contribute to a wide range of organizationally significant adverse consequences, such as loss of life, injury, damage to physical equipment, disruptions to production schedule, costly product recalls and litigation, negative publicity, steep decline in sales, regulatory sanctions, financial losses, and bankruptcy. [The information in brackets is added to distinguish between the act of a latent error and the lingering outcome of it.]

Bottom line: *Latent errors create the pre-conditions for serious events.*

Comparison of active and latent errors

Active errors trigger **events**, and you want to anticipate and prevent their occurrence during work. However, latent errors are more sinister, subtle, and enduring in their effects. Table 3.1 summarizes some of the key dissimilarities between active and latent error types.[16] The type of error in general tends to vary mostly with job function (who) and the delay in the error's manifestation (when). Based on the differences noted in the table, it should become evident that these two types of errors should be managed differently.

So, what is the bottom line of this comparison between active and latent errors? Active errors tend to drive the frequency of **events**—the occurrence, while latent errors, in contrast, leave behind latent conditions in the forms of unsafe values, process or procedural weaknesses, faulty or missing defenses, land mines, or error traps. From a management perspective, you want to anticipate and prevent (control) the occurrence of active errors in your operations, while you want to detect and correct (learn) the latent, at-risk or unsafe conditions that diminish (1) the resilience of your operations through missing or ineffective defenses and (2) the capacity of front-line workers to adapt when needed.

Table 3.1 Active vs. latent errors. A comparison of the characteristics of active and latent errors suggests that avoiding active errors tends to drive down the frequency of **events**, while finding and rectifying the results of latent errors tends to drive down the severity of **events**

Characteristics	Active errors	Latent errors
Who? (so called "perpetuators")	People in production areas: front-line workers in direct contact with hazardous processes (sharp end)	Mostly people in administration: executives, managers, engineers, clerical staff, but front-line workers can also play a part (blunt and sharp ends)
What? (result of error)	Damage, loss, or injury to tangible assets (in general): people, facilities, hardware, equipment, tools, systems, product, property, environment, etc.	Weaknesses or defects in system defenses such as facility design; equipment configuration; procedure content, accessibility, and usability; tools and resources; training program content or delivery; management policies; management systems; corporate values and norms. Tend to contribute to the creation of land mines, error traps in the workplace (latent conditions)
When? (onset of harm)	Immediate: usually faster than human beings are capable of responding to avoid harm	Later: harm realized when triggered by subsequent human activity or other triggering activity. Note: latent condition created right away; time-at-risk starts
Visible? (detectability)	Yes: observable injury, damage, or loss	No: usually hidden in plain sight (without looking for them)
Duration?	Momentary: time needed to dissipate energy, move mass, or transmit information	Long-lasting: conditions persist and accumulate until detected and corrected
Event frequency or severity?	Tend to drive the frequency of occurrence of events—the more active errors the more events	Tend to drive the severity of events that occur by reducing or inhibiting the effectiveness, presence, and robustness of defenses

Violations

As I pointed out earlier, people do not come to work intent on doing a bad job. For a variety of reasons—stability of income, pride in their work, a sense of belonging, personal integrity—most people generally want to do their jobs well. Most people care. Yet they still make mistakes. One sort of behavior needs to be examined, though, because it too can easily be made a false target for blame. I'm referring to violation—a behavior that intentionally deviates from an expectation, rule, policy, or procedure.

As you might expect by this point, the fault—even for violations—does not reside solely with the individual who committed the transgression which triggered an **event**. Well-intentioned people often circumvent policies, rules, and procedures simply because the policies, rules, and procedures don't work, and they have to find a way to get the job done, in spite of the prescribed process or expectation.

Conditions that provoke violations, at-risk choices, trade-offs, and shortcuts usually involve factors that are set in motion by the organization—its design, structure, and culture. It is the force of the blunt end of the organization (managers, designers, and administrators) pressing upon the sharp end (front-line workers) to produce certain work outputs. Here's a sample of the reasons workers have cited (with underlying causes denoted within parentheses):[17]

- Expecting that rules have to be bent to get the work done (due to unrealistic or unworkable expectations).
- Overconfidence in one's ability and experience for the job, resulting in a false sense of control (belief in having "seen it all," tradition, or a track record of few mistakes).
- Seeing opportunities that present themselves for greater efficiency or to do things "better" (affordances for less effort, and not grasping the job's "bigger picture" or fully understanding the technical reasons and risks for doing a task a certain way).
- Inadequate work planning leading to working "on the fly," solving problems as they arise, encountering surprises (misconfigurations in system or poor pre-job preparation).
- The burden of complying with requirements, or that non-compliance appears more beneficial relative to the worker's goals (a job that is physically burdensome).
- Absence of supervisory oversight (lack of oversight and minimal accountability).
- Work outcomes not traceable to individual workers (missing direct connections for personal work outcomes).
- The person's perception of risk is low, and workers lack a chronic uneasiness with regard to risk (inadequate technical knowledge and experience, or no recent history of bad outcomes).

There are two particular forms of violation that managers need to be able to differentiate: reckless and at-risk choices. A "reckless choice" is a pure gamble with the resources or assets that belong to others, while a person usually does not purpose to inflict harm. A person recognizes the risk, but knowingly takes a substantial, unjustifiable risk to accomplish an outcome that is born purely from selfish reasons. A safer, more conservative path is ignored. The act is not in error, but there is no intent to cause harm.[18] A person who makes such choices usually deserves some form of organizational sanction. A person makes an "at-risk choice" when the significance of the potential harm is not recognized, or the person feels justified in taking the risk—whether it deviates from a rule or not. The reward outweighs the risk—as seen by the person.[19] In either case, the person misunderstands the seriousness of the act. One-off at-risk choices do not deserve sanction, but coaching is definitely needed.

Whether they end up committing errors or violations, all people at the sharp end of organizations face the harsh realities of day-to-day work and its built-in obstacles. There really are no free agents in any organization—the choices people make are for the most part constrained by the system people work in (see systems thinking in Chapter 6).

What error is not

1 *Error is not sin.* As I have noted before, it is normal for human beings to err—it is part of the human condition. But, when people err, especially when an **event** involves serious injury, damage, or loss, we, as a society, tend to think there is something sinful with the person—we attribute wrong desires to the person. Error is not intentional—deliberate, premeditated, calculated, or purposeful. People do not come to work secretly wanting to hurt themselves or, much less, someone else (unless there is strong factual evidence to the contrary). People should not be ill-treated or lose respect and dignity because of error. Remember, error is simply a behavior that unintentionally deviates from a preferred behavior. In most cases, the operating procedure is the standard by which we judge error. However, procedures can also be in error, and, occasionally, not following a flawed procedure is the "preferred" behavior. Expert performance is a lot more than simply following procedures or adherence to expectations—workers must make adjustments along the way to protect **assets** from harm.

2 *Error is not an event.* Error is not a synonym for **event**. The words are often conflated, but error is not harm—injury, damage, or loss. Error is an act that triggers an uncontrolled transfer of energy, movement of mass, or transmission of information that produces harm. An **event** is an unwanted outcome involving harm to an **asset**, not the error that triggered it. However, there seems to be more interest in how the harm came to be— "Who did it?"—rather than in the harm itself. Too often, managers focus on "fault finding" rather than finding solutions. It's far more important to understand how defenses failed to protect an **asset** from harm than to assign blame to the last person who touched it.

3 *Error is not a root cause.* Errors that trigger **events** are the starting points for analysis, not the end points. Harm is the result of absent, circumvented, or ineffective defenses. Statements, such as "This **event** was due to human error," suggest that people are the single problem that must be corrected. If you think about it, **events** require teamwork and cooperation of many agents for harm to come about. The unsafe preconditions for the **event** existed long before the active error occurred (latent conditions). Accidents and **events** necessarily involve the combination / interaction of faulty or missing defenses with local triggering actions. "Error-as-cause" inhibits

learning. After identifying someone's mistake as "the **event**," analysts stop asking questions, and in-depth cause analysis stops prematurely.[20]

Basics about work context

Local factors

Human activity is generally considered to be intentional—a deliberate striving to accomplish an objective, and it occurs within a particular context—a set of circumstances in the workplace where performance occurs. *Local factors* are workplace conditions that influence a worker's behavior through multiple and varied interactions between people and the workplace.[21] *Local factors* influence people's choices on: (1) what to do, (2) how to do it, (3) when to do it, (4) where to do it, and (5) how well to do it. The word "local" has to do with the immediate geographical vicinity of the work activity, where the person is—in the here and now.

A factor is anything, past or present, that influences the person's performance here and now. *Local factors*, mostly outcomes of the organization, include all those features of the immediate workplace that are relevant to a particular person doing a specific task at a specific time and place under certain conditions. For example, what are the *local factors* influencing a recently licensed 15-year-old teenager (person) driving a 10-year-old pickup truck (task) late Saturday night (time) on a wet and curvy secondary road (place) with two or three rowdy friends with a few six-packs of beer (fitness)? You know them—it is not difficult to recognize them. An organization's goals and management systems, its facilities, the priorities and values of its managers, the quality of tools, training, procedures, and supervision all come to bear on individuals through *local factors*.

Dr. Thomas Gilbert's original research on behavior choices in business contexts identified six core aspects (categories) that blend all these workplace factors that influence performance into one model.[22] Originally referred to as the "Behavior Engineering Model," *local factors* (as I refer to them collectively) offer managers a tool for thinking about how performance is influenced—about how to "design" the system to "engineer" specific, desired behaviors. Individually, *local factors* are specific levers—drivers—that managers can use to guide or constrain people's choices. Together, *local factors* influence people's choices at the point where work is done.[23]

An organization can help its workers succeed by making sure *local factors*—the work context—are aligned to enable (encourage) the desired performance (behaviors and results) and inhibit (discourage) unwanted practices. This means influencing a performer, for a specific task, time, and place, using well-designed levers from the following categories:[24]

1 Expectations and feedback.
2 Tools, resources, and job-site conditions.
3 Incentives and disincentives (consequences).
4 Knowledge and skills.

5   Capacity and readiness.
6   Personal motives and preferences.

Figure 3.1 illustrates graphically that multiple influences impact the individual at any one time. The word "cell" refers to Table 3.2, which describes each *local factor* category.

Local factors are either external or internal to the individual. External factors are attributable to those outside the person, associated with the particular task, his/her work group, and the environmental conditions of the workplace. Internal factors are related to the generic traits of human nature we all share as human beings and to each person's unique characteristics and capabilities. Some factors are tangible while others are not. For example, people, machines, heat, procedures, money, debris, health, and tools—physical objects—are all tangible factors. But priorities, expertise, values, time pressure, motives, preferences, reputation, and relationships make up some of the intangible parts of a system. All *local factors*, internal or external to the person, tangible or intangible, work either for or against safe behavior choices. Most *local factors* are the direct outcomes, intentionally or unintentionally, of the organization and its management systems (what I collectively refer to as *organizational factors*, described in the next section and in Chapter 6).

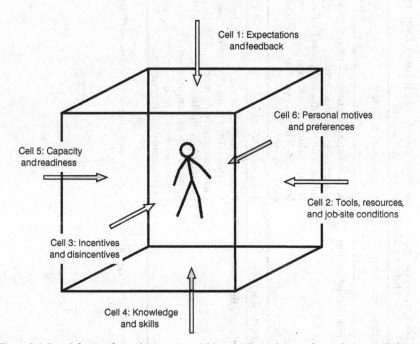

Figure 3.1 Local factors form the context within which work is performed, most of which are the outcome of various organizational functions and their management systems. An understanding of local factors provides line managers greater leverage over the choices of those who perform work

Table 3.2 Local factors categories. Originally described as the Behavior Engineering Model, these six cells, when effectively managed, can be used to "engineer" people's behavior choices within organizations

Task & environmental factors	1. Expectations & feedback	2. Tools, resources, & job-site conditions	3. Incentives & disincentives
External (outside the person)	Work-related expectations—what is acceptable and unacceptable, guidance on what one is supposed to do; coaching or correction on how well he/she performs expectations	Work-related instruments, devices, supplies, aids, and means needed for accomplishing work; external conditions in the workplace influencing actions and choices	Consequences related to adherence to and performance of expectations; rewards and sanctions, explicitly and implicitly, intended and unintended associated with the work
Individual & human factors	4. Knowledge & skills	5. Capacity & readiness	6. Personal motives & preferences
Internal (inside the person)	Person's competence and expertise related to the expectation (understanding of the work, its technology, fundamentals / theories, systems, construction), including abilities; including level of proficiency and experience	Person's physical, mental, and emotional factors influencing individual's ability / capacity to perform the expectation in the work environment	Person's attitudes, motives, personal expectations, and preferences related to his/her needs for security, achievement, affiliation, and control

Adapted from Gilbert, T. (1996). *Human Competence: Engineering Worthy Performance* (Tribute Edition). Washington, DC: International Society for Performance Improvement (p.88).

Appendix 1 provides a detailed list of common generic local factors (influencers). There is no "one-size-fits-all" approach to all jobs and workers. These influencers vary from job to job and person to person, which suggest that a conscientious manager or supervisor must thoughtfully tailor each situation to the specific people and tasks involved. A good management rule-of-thumb is to believe that *there is no such thing as a routine job*, and a manager who does his or her job well grasps the importance of tailoring these influencers to the local risks and needs of the workers in the organization. Keep in mind that modifications to the system are necessary to genuinely alter *local factors*—the direct manifestations of the organization's functions, its management systems, values, beliefs, and norms.

Error traps

What makes work hard to accomplish reliably? Human error, which can happen multiple times an hour. Avoiding or mitigating error always comes down to the specific context of the worker in which critical tasks are performed—in the here and now. Error traps are *local factors* with an unsafe attitude. Procedures that make no sense, excessive/needless documentation, hurrying, tools that don't fit or work, and domineering bosses all conspire to prevent people from getting anything accomplished reliably without error. Error traps increase the likelihood of making a mistake that enhance the likelihood of losing control. Error traps foster uncertainty or deviation potential in behavior—which in turn, creates variation in work outputs. *Variation in behavior leads to variation in results.* Error traps that provoke variation and uncertainty in workplace behavior are a strategic risk, especially during high-risk operations.

Most error traps involve a mismatch between the task and its environment and the limits of human capabilities. For example, as the speed of task performance increases, the accuracy of results tends to drop off. Error traps have been otherwise referred to as "error precursors" and "performance-shaping factors." Examples of error traps in no particular order include the following:

- hurrying
- competing goals
- high workload
- vague procedure
- distraction
- multiple, concurrent tasks
- change from routine
- schedule pressure

- unfamiliarity with a task or first time
- nearness of achieving goal (summit fever)
- interruption
- fear
- stress and fatigue
- habit
- improper tool

- inexperience or lack of proficiency with a task
- lack of knowledge or skill
- overconfidence
- unclear expectation
- out-of-service instrumentation
- confusing or no labels.

A more complete list of error traps is available in Appendix 2. It may be enlightening to a manager to query front-line workers about the more common error traps they encounter in the workplace. It will likely lead to some system changes.

Notice that the presence of error traps "tends" to increase the likelihood of error—it doesn't mean that human error is a certainty. An increase in the number of error traps tends to increase the chance for error. Decreasing the number of error traps tends to increase the likelihood the task will be performed without error. Just being a human is an error trap. There is no absolute correlation between the number of error traps and the occurrence of error, but it is more likely the preponderance of adverse influences in the workplace will provoke error.

Research associated with pilot training in the aviation community has provided great insights into error traps. Experience with crew resource management (CRM) has shown that understanding the types of factors that increase the likelihood of error enables a flight crew to remain wary for the occurrence of error. Such understanding and awareness reinforces the need for vigilance, wariness, and disciplined monitoring and cross-check strategies between pilots (especially for commercial aircraft) during critical phases of flight. Researchers concluded that a firm grasp of prevalent error traps on the flight deck can assist pilots in avoiding critical errors.[25] My own experience bears this out. Sensitizing front-line workers to the presence of specific error traps in the workplace, especially for CRITICAL STEPS, enhances their awareness of the potential for error, helping them avoid losing control.

Not all error traps are created equal. Some are far more likely to provoke error than others. Dr. Reason noted, for instance, that an unavailable or confusing procedure increases the likelihood of error 20 times compared with a similar work situation in which procedures are available and comprehendible.[26] The number 20 represents the comparative risk factor (strength) of the error trap. Below, I list a few other relevant error traps and their comparative risk factor (the likelihood of error) compared with a condition without the error trap:

- procedure not available to the worker (20×);
- worker is unfamiliar with task (17×);
- hurrying to complete a job (10×);
- worker overloaded with information (6×);
- underestimating level of risk involved in a task (4×);
- worker inexperience (3×);
- boredom or monotonous routine (1.1×).

Some error traps rank higher than others in their effects on performers in the workplace. Doesn't it make sense to warn front-line workers and their supervisors about those error traps that are particularly strong predictors of error, especially during the critical phases of work? Generally, the top three or four error traps listed in each cell of Appendix 2 are more potent than the others listed in that cell. These are provided for general guidance only and not to be used for design purposes. Be aware that there are *local factors* that exhibit a similar effect on people's likelihood to violate expectations or to make at-risk choices (see Chapter 7).

Basics about organization and management

Any outcome that surprises the organization is a failure of risk management and indicative of the organization's lack of preparation for it. In most cases, where a serious **event** occurred, managers did not fully understand how their organization worked (or did not work). One of H&OP's enduring principles is that *an organization is perfectly tuned to get the results* (and **events**) *it is getting*.

Organization creates the means of governance in accomplishing its mission. Organization and its management systems describe how managers are to conduct business at all levels, through setting priorities, developing and executing plans, managing technologies and their processes, defining and implementing programs, and monitoring and assessing performance. To achieve business results consistently, managers must understand their organization well enough to align its technologies, processes, and values that influence people's behavior choices in predictable, repeatable ways. An organization comprises several components and systems interacting in a variety of ways—but occasionally in ways unanticipated by its managers. Through the many organizational functions and management systems, various avenues of influence are created and aligned to direct or constrain people's choices in the workplace. The upper echelons of most enterprises look alike because most enterprises have similar organizational functions.

"Every accident, no matter how minor, is a failure of organization."
—K. R. Andrews
Business Professor (1953)

To effectively align the organization toward safety, it's important for us to get on the same page regarding the organization—who it is and what it does. All organizations have various generic functions performed by different people—avenues of influence. Each person has a specific role that contributes to the overall execution of the organization toward achieving its business venture.

Organizational functions—what it does

An effective organization requires the application of sound management systems that direct the behavior of employees—for the most part in repeatable and predictable ways. A "management system" is a formally established and documented set of activities, consistent with the particular needs and values of an organization, that are designed to produce specific results in a consistent manner on a sustainable basis.[27] Examples of technical management systems include work management, loss control, configuration control, operations, manufacturing, and maintenance. Examples of administrative management systems include business operations, customer relations, sales, supply chain, training, human resources, and pay and benefits. Other systems, such as the loss control, operating experience, self-assessment, and corrective and preventative action processes, help in identifying and resolving performance problems. For the most part, these systems create the repeatability and predictability managers require in the marketplace. Table 3.3 lists and describes the more common universal organizational functions.

Table 3.3 Functions of organizations. The following functions (management systems) exist in most organizations. Various organizational factors flow from these functions to create the workplace conditions (local factors) that, in turn, influence the behavior choices of the organization's members

Function	Description
Communication and information technology	Verbal and written exchange of information between organizations, individuals, and outside organizations to create understanding. Information technology and uses of various media: e-mail, signs, phone, newsletters, wall postings, intranet, meetings, radio, face-to-face, public announcement, closed-circuit television, etc.
Continuous improvement	Performance improvement processes: corrective action / preventive action (CA/PA) program, self-assessment, benchmarking, operating experience, causal analysis of **events**, trending of key indicators, change management.
Culture★	Collective and individual values, beliefs, and assumptions that influence the behavior choices of its members related to success and control of **hazards**, especially when no other guidance exists. Norms for the ways things get done. ★ Though generally not known as an organizational "function," culture is a strong emergent feature of the organization's social environment and its internal management and leadership practices.
Design	Engineering processes involved in developing functional products and processes for achieving particular aims or solving specific problems. The selection and purposing of technologies and the allocation of tasks between people and equipment. The accessibility and usability of tools, controls, equipment, work space/facilities layout, etc. (human–machine interface and human–computer interaction).
Expectations	Management's explicit descriptions of desired work accomplishments (results) and relevant behaviors, specifying what is acceptable and unacceptable, e.g., work outputs, policies, quality requirements, individual behaviors, business goals, process performance, safety requirements.
Facilities	The physical plant: infrastructure, buildings, machinery, equipment, controls, and tools used in handling **assets** and harnessing intrinsic **hazards** (energy, mass, and information) during operational processes used by personnel to achieve the organization's mission, vision, and goals. The condition, configuration, labeling, and availability of materials, tools, process systems, equipment, structures, controls, displays, and components.
Goals	Objectives people decide to pursue in achieving the organization's mission and vision—its business purposes. People at all levels must have clarity about the purposes of the organization, including safety, and how their work fits into the business. Outcomes related to production, safety, finances, market position, customer quality requirements, social standing, regulatory margin, among others. The priorities important in achieving its mission.
Hazard and loss control	Risk-management processes to protect people, property, and environment from harm, e.g., Environmental, Health, and Safety (EHS), foreign material exclusion, and Lock Out-Tag Out (LOTO), operational excellence, process safety management, Quality (Six Sigma) programs.

Function	Description
Housekeeping	Keeping work site clean and tidy. Workplace order and organization (e.g., Lean 5S systems). Foreign Material Exclusion (FME).
Human resources	Processes for recruiting, selecting, hiring, developing, promoting, and terminating personnel. Control of turnover, staffing levels, overtime standards and practices, pay and benefits. Disciplinary policies and accountability practices. Labor relations.
Maintenance and construction	Processes for building, installing, testing, modifying, and upkeep of the physical plant, systems, property, and information technology infrastructure in support of operations.
Management and leadership	Approach or style in accomplishing the organization's mission, how it applies basic management (e.g., Plan, Do, Check, and Adjust), including policy-making, planning, scheduling, and budgeting, project management, resourcing, purchasing, reviews, approvals, etc. Practices that influence the values and beliefs (culture) of its members: modeling, level of concern, reinforcing, coaching, and correcting; pride and ownership. Importance of interpersonal relationships (dignity, respect, honesty, fairness). Personal employee experiences with supervisors, managers, and executives. Includes accountability, rewards and recognition, incentives and disincentives.
Mission planning	The processes of setting the purposes or results the organization devotes resources to during its attempts to achieve its mission, vision, and goals. Direction and guidance for action regarding particular goals or issues relevant to the organization's mission and its business and regulatory environments, including its strategy, vision, plans, priorities, metrics, policies, rewards, and budget.
Operations and work management	Division of labor (roles and specialties) and coordination of effort, e.g., work functions and task assignments. Work management processes, e.g., troubleshooting, equipment status control, work planning, scoping, and scheduling. Provision of tools and resources. The *Work Execution Process*: preparation, execution, and learning.
Procedures	The sequence, manner, and methods of doing the work of the organization to control hazardous processes and to maintain system configuration. Direction for operational activities that exceed skill-of-the-craft. Documented means in accomplishing goals involving two or more organizational units.
Training	Content and delivery of training of employees regarding technical knowledge and skills needed to perform assigned work safely and effectively. Specification of competency and proficiency requirements. On-the-job training (OJT). Refresher training (ongoing). Use of a Systematic Approach to Training (SAT).

Adapted from Groeneweg, J. (2002). *Controlling the Controllable: Preventing Business Upsets* (5th ed.). Global Safety Group (pp.225-230).

Each of the functions in the table are usually implemented by one or more management systems—a framework of policies, processes, procedures, and standard practices to ensure that the organization fulfills all the tasks of that function—through the division of labor and coordination of effort.

Although listed in Table 3.3, culture is not necessarily an organizational function, but it is an important outcome of the values, beliefs, and assumptions of the management team and the workforce, and it should not be left to chance to develop unchecked. Culture describes the set of shared assumptions held by a group of people that influence how they perceive, think, and feel toward their work, including its **assets** and their operational **hazards**.[28] Ultimately, culture influences what one must do to be successful—or to survive, which offers another explanation why *work-as-done* will likely differ from *work-as-imagined*.[29] Often, front-line workers must adjust to workplace realities and surprises to accomplish their work. The culture—the prevailing values and norms of their work group and what their bosses pay attention to and reward—often governs what adjustments will be made relative to management's safety or production priorities.

Organizational roles—who it is

Most technical organizations comprise the following populations: front-line workers, supervisors, technical and administrative staff, managers, and executives. Because organizations are made up of people, the agents of the organization, culture is also part of the equation.

1   *Front-line worker*. Workers include those individuals who can directly alter, or "touch," the organization's physical **assets** during operations—especially those **assets** associated with accomplishing its mission and goals. The job site or workplace is where work is accomplished, which involves the creation of **pathways** between intrinsic **hazards** and **assets** for the purpose of creating value. Clerical and administrative workers include those who work with organizational and management processes not directly associated with operations, typically administrative in nature.

2   *Supervisor*. A supervisor is the first level of oversight—considered to be a member of the management team. First-line supervisors are usually an exempt employee who supervises front-line workers or individual contributors. For this reason, they are sometimes alternatively referred to as first-line managers. Occasionally, a company may have members of a bargaining unit (union) in foreman roles, which encompasses both supervisory and worker roles.

3   *Technical and administrative staff*. In most cases, engineers or specialists comprise the technical staff—knowledge workers. They design, install, test, and modify the technologies associated with operational processes. Additionally, there are a large number of administrative and clerical tasks—such as supply chain management, sales, customer relations, record keeping, scheduling, and communications—to be performed.

4   *Line manager*. This is a person who directs and oversees an operational group that contributes directly to the output of the organization's product or service. The line manager is responsible for meeting corporate objectives in a specific functional area or line of business by executing various administrative activities such as policy-making, target-setting, planning, budgeting, monitoring, and decision-making.

5   *Executives*. As senior managers and officers of the company, e.g., director, vice president, chief executive officer, these individuals occupy positions of significant authority over people and other resources. Executives decide on the strategic direction, markets, products, and goals, and monitor return-on-investments for their organization(s). They represent the organization and are accountable to external stakeholders for the organization's performance.

Collectively, an organization's people and the functions they perform at the various levels drive the enterprise's performance. If weaknesses persist at the organizational level, efforts to improve productivity and safety in the workplace will be unfruitful. As the following **event** clearly illustrates, **events** are fundamentally failures of organization and its management, not people.

Near hit: big pump, little pump

Sam yelled to Matt, "How are we supposed to know which breaker to shut off? There's no label."

"So, what else is new?" In his five years as an operator at Metalworks, Matt had become cynical over the lack of care in such things as labeling circuit breakers. "We'll just have to trace the lines to see which one goes to P19A." This had become a regular practice—to tag out the equipment.

The workaround made Sam nervous, having been on the job for only six months, but he respected the judgment of his more seasoned co-worker. Matt and Sam had received their work assignment at 7:00 a.m., just an hour before their shift change. The hike to the north end of the turbine building would eat up a quarter of that time, but the power to sump pump P19A had to be disconnected so the day crew could service it.

Matt groaned as the two men opened the door to trace the outside power leads from the breaker box to the pumps. The overnight drizzle had turned into a downpour. All the more reason to get this over with as fast as possible, Sam thought as he stepped out into the heavy rain.

The two pumps sat side-by-side, but P19A was noticeably larger than its companion, P19B.

Matt eyed the pair of pumps. "Whaddaya bet: P19A goes with the larger breaker inside?"

Sam nodded. "I guess. Let's follow the cables and see."

The power cables were visible all along the back of the building—except a short stretch where both disappeared beneath a metal cover. The purpose of the cover was not clear, but rather than remove it in the rain, Matt and Sam figured there would be no particular reason why the cables should cross out of sight. Just to be sure, though, they each eyed the entry and exit points as closely as possible. As an extra precaution, Matt jiggled the cable from pump P19A at the point it disappeared under the metal cover. As expected, the top cable moved at the other end when Matt did the test. Both men agreed it must be the correct cable—and that the cables did not cross unseen beneath the cover.

They were wrong.

Later that morning as sump pump P19A started abruptly, several men on the day-shift crew working on the pump shouted obscenities at whoever might have tagged out its circuit breaker. Although no one was injured in the mishap, the **event** was a serious and potentially life-threatening near hit.

The true story (names changed) of "Big pump, little pump" demonstrates an abundance of adverse latent conditions and faulty defenses. A laidback corporate culture neglected the importance of equipment labeling. The uncaring environment fostered cynicism among workers. Since the workers did not feel free to question one another—especially less experienced workers questioning "older, wiser" workers—shortcutting had become the norm. The presence of such system weaknesses meant the system was bound to fail at some point. Fortunately for the day shift, no one was injured.

The **event** described here illustrates the effect organization has on people's choices and on its defenses. Latent conditions as a result of weak *organizational factors* can inhibit the effectiveness of defenses. Misalignments in the organization can have calamitous effects on the organization's daily operations and its safety.

Alignment

As mentioned at the beginning of this chapter, human performance is never simple, involving complex interactions between people, technology, processes, etc. All relationships must be coordinated so as to work toward the organization's common good—not to work at cross-purposes. Each worker's effectiveness depends in large part on how he or she fits into the system and how its various components and processes influence the worker's choices and his/her work outputs.[30]

Every musical performance begins with tuning. Although concert-goers pay good money to hear coordinated musical sounds, they all understand the need for pre-concert chaos, while each musician in the orchestra tunes up. When an orchestra is in tune, the music flows harmoniously. That is an obvious truth for an orchestra, and although it is less apparent, it is equally true in other organizations. Collectively, not unlike an orchestra, the various organizational functions, its management systems, its people, and their cultures—*organizational factors*—work together to influence the context of performance—*local factors*, which, in turn, influence people's performance.

Figure 3.2 illustrates the *Alignment Model*. It provides a perspective on organizational performance that helps the manager better understand gaps between *work-as-imagined* (in the manager's mind) and *work-as-done* (by the front-line worker). By now, you should realize this is done by assembling influencers (*local factors*) through *organizational factors* that influence the behavior choices

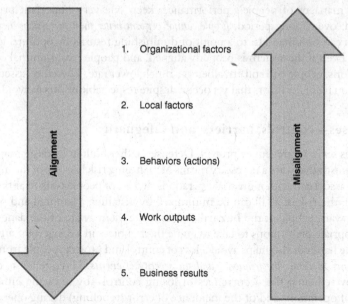

1. Organizational factors

2. Local factors

3. Behaviors (actions)

4. Work outputs

5. Business results

Alignment

Misalignment

Figure 3.2 The Alignment Model illustrates how business results are ultimately the outcome of factors originating at a system level, the domain of management. Also, it suggests the mentality an analyst should have to analyze the causes for **events**—exploring misalignments that contribute to negative outcomes

by those at the organization's operational (sharp) end. Every organization is tuned (aligned) to get precisely the results it is getting—good or bad. Preferably, performance emerges from the interaction of system components and functions in such a way that it accomplishes operations safely, effectively, and efficiently. All components of the system must work together under the direction of a driver— you, the line manager. But the effectiveness of achieving the operational and business goals depends on how those parts relate and interact with one another.

Alignment involves arranging a variety of *local factors* to get the performance (behaviors and work outputs) out of individual contributors to achieve the organization's business objectives. This suggests that managers understand which *organizational factors* influence or create specific *local factors*. For example, it is common knowledge that formal training programs develop knowledge and skills, which influence behavior choices in the workplace. Effectively aligning organizational factors to establish the proper combination of *local factors* is an engineering endeavor (some say an art). However, managers tend to be preferential in their use of *local factors*. The pet *local factors* that managers tend to rely on most are: (1) procedures, (2) knowledge and skill (training), (3) supervision, and (4) sanctions. Although these are very important factors relative to **Hu**, this is an overly simplified approach to managing the broad range of human performance issues. Reliance on such a limited set of interventions does not reflect systems thinking, which is addressed in Chapter 6.

As a manager of people's performance, keep one very important principle in mind: over a long period, *people cannot perform better than the system they work in*.[31] You can optimize **Hu** to a point of diminishing returns by optimizing your system. Even if the system is properly aligned, and people have minimal outside distractions, people will still err, albeit at a much lower rate. Therefore, assets need protection from the errors that yet occur, despite respectable efforts to avoid them.

Defenses—controls, barriers, and safeguards

Defenses serve to prevent or protect. Defenses, either built in by design or added on administratively, are a necessary means of managing risk because of the intrinsic dangers associated with industrial operations. If a lot of people walk near the edge of a cliff, the risk of a fall can be minimized by installing a handrail and posting signs to warn people of the **hazard**—signs make them aware of the danger and of appropriate precautions to take to guide their choices in a dangerous situation, while the handrail also helps avoid a loss of control and protects people from falls.

*Defenses moderate the frequency and the severity of **events***.[32] Frequency is driven down by reducing the occurrences of losing control—by reducing either the tempo of operations and/or the incidence of error (touching) during operations. Assuming a steady tempo of work, defenses that reduce (preferably, prevent) the occurrence of error are needed—*controls* (such as procedures, signs, and **Hu** tools (non-technical skills)). Severity is minimized by reducing the consequences of harm if control is lost. Defenses protect **assets** from uncontrolled **hazards**—*barriers* (such as machine guards, fences, curtains, interlocks, passwords). If barriers fail, then harm is likely. Therefore, an additional line of defense is needed to mitigate the degree of harm sustained by **assets**—*safeguards* (such as fire suppression sprinklers, first aid, escape, EMS, emergency first responders).

Whenever a serious **event** occurs, all three forms of defense usually must fail to sustain serious harm. Controls help individuals prevent a loss of control. If control is lost, barriers should work to impede the flow of energy or the movement of mass or information to protect **assets** from harm—consequences should remain small—controllable. But, if the barriers fail or are missing, then safeguards step in to minimize the harm done. But then again, if the safeguards don't work, then you have exceeded the organization's designed capabilities to protect its **assets**—you are truly out of control. The three functions of defenses and their sequence of failure for **events** is illustrated in Figure 3.3.

Figure 3.4 illustrates the application of these three types of defenses to protect a tank from damage from over-pressurization. If pressure is being controlled manually, an operator is directed by a procedure to keep pressure within a defined operating range. If the operator per chance becomes distracted, and pressure creeps up out of control, an audible alarm alerts the operator of a high-pressure condition. Alerted, the operator should regain control of pressure and return it to within the approved operating band. Both the procedure and the alarm serve as controls—guiding and alerting the operator's responses in controlling tank pressure. However, if the operator does not respond promptly and pressure

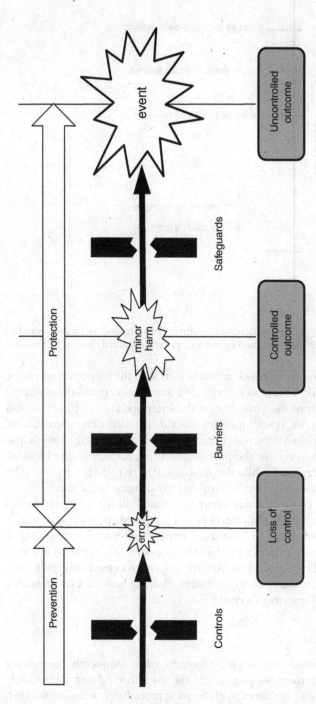

Figure 3.3 Defenses: controls, barriers, and safeguards. Controls guide the choices of front-line workers. Barriers impede the movement of energy, mass, and information. Safeguards mitigate the harm done for outcomes that exceed design assumptions. (Adapted from Figure 3.2 in Hollnagel, E. (2004). *Barriers and Accident Prevention.* Aldershot: Ashgate.)

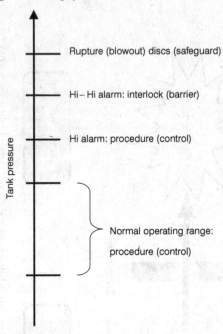

Figure 3.4 Example of overlapping defenses (controls, barriers, and safeguards) used to protect a tank from over-pressurization in order to avoid substantial damage to the tank

continues to rise, then an interlock activates to isolate the source of energy or fluid entering the tank. This is an example of a barrier in that additional energy or mass is isolated from the tank. But, if the interlock fails to function, and pressure continues to rise, reaching a value that threatens to burst open the tank, permanent damage to the tank would occur. However, an installed blowout disc ruptures as designed to relieve the pressure, limiting damage to the blowout disc, which can be replaced while the tank is out of service for repairs. The blowout disc serves as a safeguard, minimizing the damage to the tank.

As you can see, controls guide people's choices aimed at prevention— avoiding the loss of control, while barriers and safeguards provide **assets** with various forms of protection from unwanted transfers of energy, movements of mass, or transmissions of information. Each type of defense is explored in a little more detail in the following sections. Reinforcing controls tends to regulate the frequency of **events**, while restoring and strengthening barriers and safeguards tend to minimize the severity of **events**.

Controls

Controls guide the behavior choices of performers, and consequently, controls are usually integrated into the work process and the workplace.[33] Most controls rely on people's adherence to use them, such as using procedures, adhering to signs, rules, and regulations, reviews and approvals, supervisory oversight, and various expectations, such as the use of **Hu** tools—such as self-checking, peer checking, and three-part communication. The intent of controls is to avoid loss of control by preventing error (reducing the likelihood that it will happen). Controls aid in detecting unsafe conditions, or otherwise informing, creating awareness, and discouraging unsafe acts. Consequently, controls tend to reduce the occurrence of active errors, thus minimizing the probability or frequency of **events**.[34]

Consider railroad grade crossings at local streets and roads in populated municipal areas. Every example listed below requires people to acknowledge and heed them for their own protection. Common controls encountered at such crossings include:

- crossbuck (X-shaped) railroad road sign (alerts vehicle drivers);
- crossbuck sign painted on surface of street or road (alerts vehicle drivers);
- flashing red lights (alerts vehicle drivers to an approaching train);
- warning bell (alerts vehicle drivers and pedestrians);
- train horn (alerts vehicle drivers and pedestrians).

The central weakness associated with controls is that their effectiveness depends entirely upon people—their willingness or remembrance to abide by them. Therefore, a control's reliability is generally consistent with nominal human reliability ($\approx$99 percent, depending on circumstances).

Barriers

Barriers are designed to block the transport (or interruption) of energy, mass, or information between points in a system. Additionally, barriers may restrict the free movement of objects. Barriers either prevent release or entry, depending on what is being protected. They offer **assets** protection against intrinsic **hazards** and things people unintentionally or unwittingly do wrong. Personal protective equipment (PPE), such as hard hats, gloves, and safety shoes, keep dangerous objects from penetrating a worker's body; the "undo" feature in word processing software offers the opportunity to reverse an inadvertent keystroke; machine guards prevent inadvertent contact with rotating machinery; passwords impede unauthorized access to sensitive information; and locks on equipment controls avoid unintended operation during maintenance.

Returning to the railroad-crossing example, approximately 20 seconds before a train arrives, a gate automatically lowers across the street to inhibit vehicular traffic from crossing the grade while the train approaches or occupies the crossing. Gates remain lowered until the train clears the street. The gate serves as a barrier, impeding the flow of vehicles through the crossing.

Except for passive, built-in defenses—such as walls—the primary source of a barrier's weakness is related to the quality of its construction and the occurrence of regular maintenance, which again depends on people and organizations.[35] Barriers are more reliable than controls but depend on management systems that confirm or certify the presence, integrity, and robustness of the barriers. From time to time, people will purposefully circumvent barriers for various reasons, some legitimate, some illegitimate.

Safeguards

Safeguards minimize the severity of **events** that still occur, despite the controls and barriers that may have been in force. Safeguards mitigate harm, provide a way of escape, and improve the chances of recovery. Fire suppression sprinkler systems in commercial buildings, emergency medical services, and crash bars on doors at building exits are examples of safeguards. You'll note that safeguards come into play after something has gone wrong (controls and barriers failed) and the containment of injury, damage, or loss is now required. A fire has started, a traffic accident has happened, or a toxic gas release has occurred that requires rapid evacuation from a facility. For each, you are beyond the point of keeping the problem from starting. Some harm has occurred, and now one or more safeguards are needed to help us out. Safeguards minimize destruction, limit the toll an injury takes on a victim, or allow people to flee to a safe place to avoid injury or death.

Safeguards at railroad crossings generally do not exist, depending on the availability of emergency responders. If there is a collision, someone at the scene usually calls 911 to summon the local Emergency Medical Service (EMS) to render aid to any casualties and the local fire department to extinguish any fires or control the release or spill of hazardous materials, if any. In either case, their aim is to limit the degree of injury and damage to people and property.

In some high-hazard operations, defense-in-depth is necessary. It is achieved by arranging multiple or redundant defenses in an overlapping fashion to protect people, the plant, and other assets from danger, such that a failure of one defense would be compensated for by one or more other defenses, avoiding harm.[36] Table 3.4 summarizes the three types of defenses, providing examples of each.

Land mines

Pulling the trigger on an empty firearm will not hurt anyone or damage anything. But if a bullet is unintentionally left in the chamber, pulling the trigger on a gun thought to be empty can kill someone or destroy something. An unknown source of harm (bullet in chamber), or a missing defense (malfunctioning safety) are land mines in any operation.

Land mines comprise undetected adverse conditions in the workplace that are poised (**pathways** exist) to trigger harm—workplace conditions that increase the potential for an uncontrolled transfer of energy, mass, or information with one action—an unexpected source of harm—unbeknownst to the performer. Land mines usually exist because of degraded or missing defenses. These are "accidents waiting to happen." For example, a mispositioned manual valve—left open during a previous work activity—creates the potential for an unwanted transfer of fluid during subsequent evolutions that presume that same valve is closed.

Table 3.4 The three types of defenses: controls, barriers, and safeguards, with examples of each

	Defenses	
Controls	*Barriers*	*Safeguards*
Measures that guide, coordinate, or regulate the behavior choices of people	Measures that protect assets against harm by limiting or impeding the free movement or flow of energy, objects, substances, or information	Measures that mitigate the onset of harm or protect the integrity and security of assets against large-scale, widespread harm to people, facilities, organizations, the environment, and surrounding communities
	Examples	
• knowledge and skills • caution statements in procedures • checklists • policies, rules, expectations • signs and signals • **Hu** tools (self-checking, peer-checking, pre-job briefing, etc.) • values, beliefs, and principles • group norms • process indicators (gages) • supervisory monitoring • alarms • rumble strips (highways) • field observations by managers • trending • caution tags • lines and color coding	• interlocks (digital, electrical, and mechanical) • physical lock and key • firewall (software) • railings, walls, and fences • safety belts and air bags (autos) • machine guards • railroad crossing barriers • valves and ventilation dampers • tank berms • Personal Protective Equipment (PPE): gloves, hard hat, face shield, safety shoes, eyewear, ear plugs, life jacket • electric wall socket covers • access control to facilities • passwords and fingerprints • work authorization or permit to work • speed bumps • quality control inspections	• relief valves • first-aid kits • fire extinguishers • Emergency Medical Service (EMS) • fire suppression sprinklers • spares • blowout discs on tanks • computer file or server backup • emergency eye-wash stations (for chemical hazards) • firefighting equipment • emergency control center • emergency cooling systems • security force • emergency fund (money) • Automated External Defibrillators (AED) • satellite telephones • plan "B"

Compiled from information available in Hollnagel, E. (2004). *Barriers and Accident Prevention.* Aldershot: Ashgate.

From a bucket of paint perched on an unstable stepladder to the handle of a skillet extending over the edge of a stove top—accessible to a toddler—we all recognize certain hazardous situations. Although land mines in a workplace are less obvious than day-to-day examples, a healthy organization will be diligent to look for them. A search will turn up land mines such as:

- pressurized systems;
- mispositioned components;
- out-of-service indicators;
- inoperable emergency systems;
- long-term equipment deficiencies;
- inaccessible controls;
- missing or weak barriers (e.g., guards and railings);
- worn-out or missing tools;
- hidden configurations;
- software bugs.

None of the above are unusual or out of the ordinary. A key to understanding and rooting out land mines is to acknowledge that they "aren't supposed to be there" but are. The wary worker knows that just because something isn't supposed to be a certain way doesn't mean that it isn't. There can even remain land mines after being discovered if prompt corrective actions have not been taken to make the condition safe.

The effectiveness and quality of defenses are swayed mostly by their respective management systems that governed their design, construction, and subsequent preservation. Management systems influence the presence and effectiveness of defenses in a variety of ways such as the following:[37]

- robustness (ability to withstand extreme **events** or environmental conditions);
- delay in implementation;
- availability when needed (reliable—will function when called upon);
- degree of difficulty of use;
- degree of dependence on human intervention;
- cost to implement and maintain.

If there are system weaknesses, there will likely be weaknesses with defenses. Detecting and correcting latent system weaknesses improves an organization's defenses and its resilience, tending to drive down the severity of **events**.

Severity pyramid—misunderstood

During the 1930s, American industrial safety expert Herbert Heinrich popularized in his 1931 book, *Industrial Accident Prevention, A Scientific Approach*, an apparent causal relationship between job-site errors and serious workplace industrial safety accidents. This relationship became known as Heinrich's law or the Heinrich ratio: there is a fixed ratio between **events** having progressively greater severity.[38] Unfortunately, this has been misunderstood by managers to mean that if the number of low-severity **events** (errors and near hits) is reduced, higher severity **events** (deaths, equipment damage, and disabling injuries) can be avoided. Reducing errors will indeed reduce the frequency of **events**, but does little to minimize the severity of any **events** that do occur. Correcting this misunderstanding is highly significant to our discussion. Going forward, I refer to this ratio as the severity pyramid (see Figure 3.5).

Some people interpret the severity pyramid to suggest that the "causes of major incidents are the same as the causes of minor **events**." This is not true. The pyramid is often misconstrued as presenting a causal relationship between its layers— between the number of errors (potential **event**-triggering actions) and the number of serious safety **events**. This leads to the erroneous belief that if an organization reduces the number of errors made, then the number of deaths or disabling injuries will likewise drop.[39] If you erroneously believe the pyramid to demonstrate a causal relationship, our attempts to decrease the severity of **events** will tend to focus only on reducing the number of errors that trigger **events** and not on the other factors that really lessen the consequences of incidents—barriers and safeguards. In truth, each level of the severity pyramid only denotes different types of **events**.[40] In this sense, the severity pyramid predicts no causal relationship between the layers of the pyramid whatsoever—hence, it is irrelevant to **event** severity.

Unwanted variations in performance—human error—do not necessarily lead to undesired outcomes. Why is it that slips, lapses, fumbles, trips, or mistakes trigger harm and damage in one case, but not so in other cases? What error reduction actually does is keep work outcomes off of the pyramid entirely—no loss of control. For the most part, where no error occurs, no **event** occurs, and this is largely attributable to effective controls, such as the persistant and

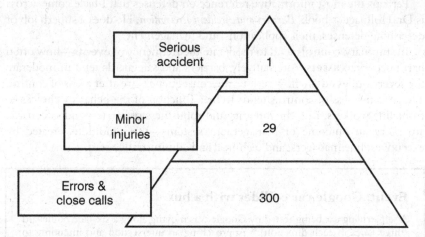

Figure 3.5 The Heinrich triangle erroneously suggests a causal relationship between each level of severity. (Source: Heinrich, H. (1931). *Industrial Accident Prevention*. New York: McGraw-Hill.)

rigorous application of **Hu** tools. Conclusion: barriers and safeguards, not error avoidance, tend to minimize the severity of **events**.

The moment error occurs, a whole new set of dynamics comes into play that determine how harsh the consequences will be. To be clear: the very error that results in no consequence in one situation (bottom level of the pyramid) could easily be the same error that kills (top level), and it could be you. What's the difference?

When an error occurs in the workplace, its outcome essentially depends on the presence, integrity, and robustness of **assets** and the defenses present—not the error that triggers a loss of control. Several defenses usually fail in an **event**—the operation and its organization encounters a scenario that surprises everyone. Most serious **events** are "caused" by system weaknesses that either (1) weaken the effectiveness of defenses or (2) allow them to be circumvented altogether. The "causes" of serious **events** are failures associated with ineffective or insufficiently robust defenses—the outcome of various system weaknesses. As noted earlier, this leads to the conclusion that all serious **events** are fundamentally organizational failures—failures of one or more organizational functions and their related management systems or insufficient risk management. A substantial weakness with Heinrich's law is that it places a disproportionate weight on the unsafe acts and errors of individuals as the cause of **events** and gives insufficient attention to factors stemming from management and organizational systems that contributed to the failure of defenses—controls, barriers, and safeguards.

"Large scale disasters need time, resources, and organization if they are to occur ..."

—Barry Turner and Nick Pidgeon
Co-authors: *Man-Made Disasters* (1978)

Perhaps the most informative reference on defenses that I have come across is Dr. Hollnagel's book, *Barriers and Accident Prevention*. He does a superb job of describing defenses, their applications, and management.

In summary, controls tend to moderate the frequency of **events**—how often harm occurs to **assets**. Alternatively, barriers and safeguards tend to moderate the severity of **events**—how serious the consequences are after a loss of control. The effectiveness of controls tends to be a function of the behavior choices of front-line workers. But, the integrity and robustness of barriers and safeguards are more an outcome of management systems—the conditions created by executives, line managers, and technical and administrative staff.

Event: Google car collides with a bus

A self-driving car being tested by Google was moving along a six-lane boulevard (three lanes in each direction).[41] Approaching an intersection and intending to make a right turn, the car was in the right part of the lane. The lane was wide enough for both right-turning vehicles and others that were proceeding straight through the intersection. However, the Google car software detected an obstacle ahead—sand bags protecting a damaged storm drain. Apparently, the Google car stopped momentarily before moving to the left in the same lane at 2 miles per hour (mph). Both the car's "digital control system" and its passenger (supervisor/driver) were aware of a bus approaching from behind, and both assumed that the bus would yield. The bus driver saw the car but assumed it would yield to him. Neither driver yielded. The Google car collided with the bus, and a "fender-bender" occurred. No one was injured.

The Google car's passenger (supervisor) "could have" overridden the software, taking control of the car via the steering wheel to avoid the bus (adaptive behavior). By law, the bus driver "should have" yielded (driving-as-imagined). Driving in the "real world" does not always observe the letter of the law. Google stated that the collision is "a classic example of the negotiation that's a normal part of driving—we're all trying to predict each other's movements" in a complex environment! Both operators could have avoided the incident, but most of us blame the deficient software.

The software controlling the Google car was "designed" to cope with traffic patterns and normal (human) behavior—driving-as-imagined. The drivers of buses, large trucks and other large vehicles, sometimes intimidate or assert their right of way, because of their size, limited maneuverability, weight, and longer stopping distances. These drivers expect drivers of smaller vehicles to respect them (driving-as-done). But, as you see here, software engineers did not and *cannot* anticipate all situations, scenarios, or conditions that a new technology will eventually encounter in the real world.

The incident described here offers several learning opportunities. However, the one I want to emphasize is that the workplace—the design of its technologies and the content of its procedures—is fraught with assumptions about how work is supposed to happen (*work-as-imagined*). This is idealistic. There are hidden conditions in the workplace that front-line workers and operators must recognize and respond to in order to maintain safety during work despite the presence of automation and other built-in defenses. To preserve safety, they must be able to cope with the nuances and surprises that often pop up in the workplace. If either driver had exercised caution and adapted appropriately, taking conservative action, the accident would have been avoided. In the real world, *work-as-done* often differs from *work-as-imagined* for this reason. In most cases, there are no procedures for these occasions.

Resilience—adaptive capacity of front-line workers

Things do not always go as planned. Things wear out. People are fallible. There is always uncertainty in the workplace. What will workers do when they are surprised, or the demands and tempo of the work suddenly escalate? What if scheduled production activities inadvertently put **assets** in harm's way? What if a new work situation arises that does not match the procedure or contingency plans? Customarily, the first thing that occurs in any **event** or accident is a disturbance, disruption, or someone or something deviates from the plan and its assumptions (*work-as-imagined*). Front-line workers encounter disruptions regularly. Occasionally, front-line workers and their supervisors encounter workplace situations that:[42]

- exceed their capacity, authority, or on-hand resources—things changing too fast or not enough people or raw material;
- involve competing goals or goals that work at cross-purposes with others— to get things done safely, on schedule, and under budget;
- have no guidance on how to respond to them—uncertain nature of the process, no procedure, inexperience, novel situation, or lack of expertise.

Dr. Erik Hollnagel, in his book, *Safety-I and Safety-II*, summarizes resilience as the "ability to succeed under varying conditions."[43] Resilience means you can experience shocks and survive—you can take a hit and keep moving, without losing control. Managers and engineers do their best to design and build robust defenses of various forms and functions into the operation so as to accomplish work in the context of intrinsic operational **hazards** and changing conditions. Beth Lay, author of several resilience engineering articles and chapters, says, "Not everything can be known, and not everything can be predicted." Consequently, it is important to continuously monitor high-risk operations to make real-time adjustments—adaptations—to changing risk exposures throughout the work day.

How do workers respond rather than simply react to avoid harm or disruption? What do workers do? When procedures do not work, front-line workers should stop and ask for help. But, if time is unavailable, they will have to adapt, adjusting their activities on the fly. Do people know how to make technically competent field adjustments? Do they have the prerequisite technical expertise to make conservative decisions? A conservative decision places the safety of **assets** above production goals (see Chapter 4 and Appendix 5).

Adaptive capacity involves the front-line personnel's ability to (1) realize when reliable technical guidance for a particular work situation is unavailable and (2) make conservative decisions that protect **assets** from impending harm. Adaptive capacity promotes flexibility, innovation, and creativity and enables people to anticipate—to "think outside the box"—and adjust current plans—to "shift on the fly"—to avoid harm during work. Yet, such an approach creates another kind of uncertainty that can be unnerving to many line managers— giving front-line workers the authority (control) to make field adjustments to protect **assets**. This improves an organization's resilience—its capacity to cope with the unexpected and still operate safely. Improving the adaptive capacity of front-line personnel improves the dependability of their adjustments to dynamic risks in the workplace—especially the unexpected. Consequently, there is less pressure on crafting a "perfect" procedure and more dependence on the "thinking operator." Adaptive capacity bridges the gap between the way things are and what they need to be for safety. In a manner of speaking, adaptive capacity—enhancing the technical soundness of field adjustments—involves improvisation, but not improvisation for production's sake—safety's sake. This is what is meant by the statement, "Safety is what you do." However, adaptive capacity depends on developing and sustaining technical expertise.

"A reliable system is one that can spot an action or function going wrong, not an action gone wrong."

—Karl Weick and Kathleen Sutcliffe
Co-authors: *Managing the Unexpected* (2015)

Building adaptive capacity into the workforce places a great deal of emphasis on developing technical expertise at the worker level. Without it, the dependability of technical adjustments in the workplace would be suspect. The expertise of front-line workers serves as a technical backup, a form of defense-in-depth, to the plans and oversight by line managers, the procedures authored by process specialists, and an engineer's designs. Although, because of human fallibility, front-line workers remain a **hazard** to **assets** during operations, they act, at the same time, as heroes, accommodating gaps and vulnerabilities in processes and defenses. Training is addressed in further detail in Chapter 8.

Matthieu Branlat and David Woods, in a conference paper on adaptive capacity, described resilience in a way that I think is particularly relevant to the workplace:[44]

From the study of work systems, resilience appears as more than simply bouncing back from disruptions, as it is often defined. Anticipating and avoiding disruptions, rather than simply reacting to them, is actually a much more powerful way to cope with the variability of the real world. In reliable organizations operating in high-risk domains, anticipation doesn't simply mean preparing contingency plans for any imaginable disruption (although this can be useful), but it rather corresponds to the belief that surprising **events** will occur and challenge the system's assumptions about the world. A resilient system therefore needs to be able to deploy new ways of functioning... As a result of these points, we define resilience as fundamentally anticipatory, and related to the capacity to handle the next disruption.

Technically competent front-line workers are the first-line of defense for high-reliable, resilience organizations. But, if people fail, which they will soon enough, there is need for backup defenses. Defense-in-depth builds resiliency into an organization—a characteristic that allows an organization to suffer little or no ill-effects from a loss of control. But defense-in-depth has a downside in that it increases complexity and sometimes makes it difficult for operators to fully understand what is really going on. To accommodate less operational transparency created by defenses-in-depth, a resilient organization has an additional defensive characteristic in its workforce—the ability of front-line workers to adapt, responding safely to unexpected situations. This flexibility—the capacity to detect and respond to situations that pose threats to **assets**—is manifested through the organization's people and provides the best source of safety for the unexpected.

You may not anticipate every harmful scenario that could be encountered in the workplace, but you can be ready for anything. In the next chapter, RISK-BASED THINKING is introduced—which serves as a foundation for safety and resilience in operations throughout the organization.

Things you can do tomorrow

1   As part of work planning, operations, and risk-management meetings, use the vocabulary of H&OP. Target specific words and phrases—relevant to the work at hand—include concepts in conversations during regular meetings.

2   Review a sampling of **event** reports from the last two to three years that concluded that human error was the "root cause." Identify the controls, barriers, and safeguards that failed and their related management systems. Identify repetitive events, active errors, error traps, and recurring management system weaknesses.

3   Review the detailed list of error traps in Appendix 2, translate them into objective, yes/no questions that front-line workers can use to assess the error trap's presence and influence (verifiable) on the tasks at hand. During refresher training or at other venues with front-line workers, ask them to identify the top 10 workplace error traps for their specific workgroup or for recurring high-risk operations. Ask them how each error trap could either be eliminated or minimized.

4   Use the *Alignment Model* to review the next **event** discussed by the management team. Explore the factors influencing people's choices as well as the reasons defenses did not work as advertised.

5   Incorporate H&OP fundamentals in line training programs. Include the H&OP glossary into initial and refresher technical training for front-line workers and other operational positions.

6   Train operational personnel on how to make conservative decisions in the field, considering various scenarios that could arise. During a pre-shift briefing, present a work-related scenario that poses goal conflicts and pits production goals against safety goals. Briefly talk about the technical problem and the risks to **assets** and what to do to ensure their safety.

Notes

1   Kletz, T. (2001). *An Engineer's View of Human Error*. (3rd edn). Boca Raton, FL: CRC Press (p.2).
2   Hofman, D. and Frese, M. (Eds.) (2011). *Errors in Organizations*. New York: Routledge (p.1).
3   Marx, D. (2015). *Dave's Subs*. Plano, TX: By Your Side Studios (p.251).
4   Conklin, T. (2012). *Pre-Accident Investigations: An Introduction to Organizational Safety*. Farnham: Ashgate (p.69).
5   This sentence sounds like a profound quote that should be attributed to someone, but I can't find it. However, I attribute the principle to the work of Dr. Elliot Jacques in his book, *Requisite Organization* (p.6).

6 Reason, J. (2008). *The Human Contribution: Unsafe Acts, Accidents and Heroic Recoveries*. Farnham: Ashgate (p.34).

7 Hollnagel, E. (1998). *Cognitive Reliability and Error Analysis Method: CREAM*. New York: Elsevier (p.251).

8 Reason, J. (2013). *A Life in Error: From Little Slips to Big Disasters*. Farnham: Ashgate (pp.7–12).

9 Civil Aviation Safety Authority of Australia. (2013). *Safety Behaviors Human Factors: Resource Guide for Engineers*. Canberra, Australia: Civil Aviation Safety Authority (p.27). Also, Dr. Michael Frese of the University of Giessen and London Business School suggests that you make approximately 3 to 4 errors per hour on every task that you work on, the "Frese Law of Error Frequency" as mentioned in his presentation, "Learning from Errors by Individuals and Organizations," to the American Psychological Society in Chicago, May 2008.

10 Kletz, T. (2001). *An Engineer's View of Human Error* (3rd edn). Boca Raton, FL, USA: CRC Press (p.145).

11 Reason, J. (1990). *Human Error*. New York: Cambridge (p.173).

12 Center for Chemical Process Safety (1994). *Guidelines for Preventing Human Error in Process Safety*. New York: American Institute of Chemical Engineers (pp.41–42).

13 Reason, J. (1990). *Human Error*. New York: Cambridge (p.203).

14 Reason, J. (1997). *Managing the Risks of Organizational Accidents*. Aldershot: Ashgate (p.10).

15 Ramanujam, R. and Goodman, P. (2011). The Link Between Organizational Errors and Adverse Consequences: The Role of Error-Correcting and Error-Amplifying Feedback Processes, In Hofmann, D. and Frese, M. *Errors in Organizations*. New York: Taylor and Francis (p.245).

16 Reason, J. (1997). *Managing the Risks of Organizational Accidents*. Aldershot: Ashgate (p.11).

17 Most of the reasons listed are cited in Hudson, P.T.W., et al. (2000). Bending the Rules: Managing Violation in the Workplace, *Exploration and Production Newsletter*, EP2000-7001 (pp. 42–44). The Hague: Shell International.

18 Marx, D. (2015). *Dave's Subs: A Novel Story About Workplace Accountability*. Plano, TX: By Your Side Studios (pp.106–110).

19 Ibid. (pp.151–156).

20 Woods, D., Dekker, S., Cook, R. and Johannesen, L. (2010). *Behind Human Error*. (2nd edn). Farnham: Ashgate (pp. 235–237).

21 Reason, J. (1997). *Managing the Risks of Organizational Accidents*. Aldershot: Ashgate (p.141).

22 Adapted from Gilbert, T. (1996). *Human Competence, Engineering Worthy Performance*, International Society for Performance Improvement (ISPI) (pp.82–89).

23 Van Tiem, D., Moseley, J.L. and Dessinger, J.C. (2004). *Fundamentals of Performance Technology: A Guide to Improving People, Process, and Performance* (2nd edn). Silver Springs, MD: International Society for Performance Improvement (pp. 8–10).

24 Binder, C. (July/August 1998). The Six Boxes™: A Descendent of Gilbert's Behavior Engineering Model. *Performance Improvement*. Vol. 37, No. 6 (pp.48–52). Retrieved from http://www.sixboxes.com/_customelements/uploadedResources/SixBoxes.pdf. Tom Gilbert developed the Behavior Engineering Model (BEM), based on his experiences. It was derived originally from B.F. Skinner's behavior science. Carl Binder (1998), beginning with the BEM, adjusted Gilbert's language to create the Six Boxes® Model, including numbered cells. The headings of this model of *local factors* are based on the works of Gilbert and Binder, adopting some of their language but altering some to better communicate with the reader.

25 Thomas, M. (2004). Error Management Training: Defining Best Practices, *ATSB Aviation Safety Research Grant Scheme Project* 2004/0050 (p.13).

26 Reason, J. (1997). *Managing the Risks of Organizational Accidents*. Aldershot: Ashgate (pp.142–147).

27   Center for Chemical Process Safety (2007). *Guidelines for Risk-Based Process Safety*. Hoboken, NJ: Wiley (p.10).

28   Schein, E. (1992). *Organizational Culture and Leadership*. New York: Jossey-Bass (p.12).

29   Hollnagel, E. (2014). *Safety-I and Safety-II: The Past and Future of Safety Management*. Farnham: Ashgate (p. 118).

30   Rummler, G. and Brache, A. (1995). *Improving Performance: How to Manage the White Space on the Organization Chart* (2nd edn). San Francisco, CA: Jossey-Bass (p.64).

31   Deming, W. (1986). *Out of the Crisis*. Cambridge, MA: MIT (p.315).

32   Viner, E. (2015). *Operational Risk Control: Predicting and Preventing the Unwanted*. Farnham: Gower (p.103).

33   Controls as used here is somewhat different than that used in traditional occupational safety and health terminology. In occupational safety terminology, the concept of the hierarchy of controls classifies the various types of controls in terms of their ability to prevent harm to people. The term controls, as used in occupational safety, is synonymous with the term defenses as I use it here. I acknowledge the insight of Ms. Dee Woodhull for pointing out this distinction.

34   Hollnagel, E. (2004). *Barriers and Accident Prevention*. Aldershot: Ashgate (p.80), and Trost, W. and Nertney, R. (1995). *Barrier Analysis* (DOE-01-TRAC-29-95). Boulder, CO: Scientech, Inc. (p.10).

35   Hollnagel, E. (2004). *Barriers and Accident Prevention*. Aldershot: Ashgate (p.100).

36   Ibid. (pp.70–72).

37   Ibid. (pp.97–108).

38   Heinrich, H. (1931). *Industrial Accident Prevention: A Scientific Approach*. New York: McGraw-Hill, quoted in Hollnagel, E. (2009). *Safer Complex Industrial Environments: A Human Factors Approach*. Boca Raton, FL: CRC Press.

39   Hollnagel, E. (2014). *Safety-I and Safety-II: The Past and Future of Safety Management*. Farnham: Ashgate (p. 71).

40   Ibid. (pp.73–74).

41   Pritchard, J. (2016, February 29). Google self-driving car strikes bus on California street. *Associated Press*. Retrieved from http://bigstory.ap.org/article/4d764f7fd24e4b0b9164d08a41586d60/google-self-driving-car-strikes-public-bus-california

42   Branlat, M. and Woods, D. (2010). How do Systems Manage Their Adaptive Capacity to Successfully Handle Disruptions? A Resilience Engineering Perspective. Presented at AAAI Fall Symposium on Complex Adaptive Systems—Resilience, Robustness, and Evolvability (FS-10-03). Arlington, VA, November 11–13, 2010: Association for the Advancement of Artificial Intelligence (pp.26–34).

43   Hollnagel, E. (2014). *Safety-I and Safety-II: The Past and Future of Safety Management*. Farnham: Ashgate (p. 137).

44   Branlat, M. and Woods, D. (November 2010). "How do Systems Manage Their Adaptive Capacity to Successfully Handle Disruptions? A Resilience Engineering Perspective." from Complex Adaptive Systems—Resilience, Robustness, and Evolvability: Papers from the AAAI Fall Symposium (FS-10-03) (p.30).

4 RISK-BASED THINKING and chronic uneasiness

A system is resilient if it can adjust its functioning prior to, during, or following events (changes, disturbances, and opportunities), and thereby sustain required operations under both expected and unexpected conditions.

Erik Hollnagel[1]

Only people can hold together these inherent imperfections, who create safety through practice at all levels of an organization.

Sidney Dekker[2]

Nothing is always as it seems

System operations are rarely trouble free. Both conditions and people are constantly changing. Planners cannot and do not anticipate all contingencies or situations. Budgets are approximations. Equipment wears out. Plans are incomplete. Schedules are rough guesses, based on experience. Inaccuracies exist in most technical procedures. Systems, equipment, and components may not be properly configured. People miscommunicate and make inaccurate assessments of capabilities and resources. The commercial workplace is fraught with multiple goals and organizations competing for scarce resources—there's continual pressure to produce more for less in faster time.[3] People throughout the organization attempt to achieve their objectives—faster, better, cheaper, and safer—all inherently conflicting.

What do you want people to do when:

- current plans won't work for current conditions?
- conflicting goals arise?
- resources are inadequate for the task at hand?
- demands or tempo suddenly escalate?
- they're surprised?

While it would seem that this complicated array of considerations would raise everyone's awareness about the potential for problems to develop, there is an almost universal tendency among executives, managers, and even workers to be overly optimistic about how their systems perform—perhaps because nothing bad has happened. Yet, this attitude actually makes them increasingly likely to fail. Wisdom demands that organizations become more realistic about what it takes to create and maintain a productive and safe system. Things do not always go as planned. Yet, to improve reliability and to optimize resilience in a hazardous environment, you must recognize several unrealistic assumptions about organization. Here is a list of faulty thinking that tends to promote an overly optimistic view of organizations and systems:[4]

- You believe facility systems are well designed and scrupulously maintained—the physical plant is, in effect, safe.
- You think designers foresee and anticipate all contingencies, even minor ones—there are no surprises.
- You consider procedures to be complete and correct, available, and usable for every occasion—every job is routine.
- You assume that people will behave as we expect them to—as they are trained to.
- We presume that compliance with requirements makes us safe—we're safe.
- You assume that if we pass all evaluations and audits, improvement is unnecessary—nothing needs to change.

None of these assumptions are ever fully true. No system is perfect. In reality, operations, the workplace, and its organizational system are all fraught with uncertainty, complexity, change, limited resources, and inherent conflicts—all dynamic. And, nothing is always as it seems. The work conditions encountered by front-line workers almost always differ from what is specified in procedures and instructions.[5] Compliance is important, but it is dangerous if performed mindlessly, without thinking. Therefore, front-line workers must be able and ready to adapt when needed to protect **assets** from potential harm when they encounter unanticipated **pathways** between **hazards** and **assets**. Being able to adapt means front-line workers must possess higher levels of technical knowledge and possess a clearer sense of the organization's business purposes. Such an insight about the organization and the immediate work environment should understandably stimulate a degree of uneasiness and caution toward one's work.

Important reminder. In the **event** described below, which occurred almost 30 years ago, the mechanics were the **assets**. The **hazard** to the safety of the mechanics was a rotating shaft—the **pathway** for harm since they were working the pump. The rotation of the shaft was prevented by an open circuit breaker interrupting power to the electric motor that operated the pump. Today, such circuit breakers would normally be racked out, tagged, and locked out. Kelly was the human being at the motor's circuit breaker controls—at the **touchpoint**.

Event: first impressions are last impressions

Traffic had done its best to frustrate Kelly's commute to the plant, but the young electrician remained zealous about his new job. Kelly was very proud of his recent qualification as a journeyman electrician. After so much training, he was eager to demonstrate his newly acquired skills. Now, two hours later, he sat in a situation even more frustrating than afternoon rush hour traffic. Kelly stared at the 4000-volt (4 kV) circuit breaker he was to rack out,[6] with his right hand on the control switch.

This particular evening, Kelly's crew was shorthanded. Wes, another crew leader, assigned Kelly the task of racking out several 4 kV circuit breakers for a periodic overhaul. Kelly's normal crew leader had called in sick, and so Wes had responsibility to oversee two crews. Wes knew Kelly was recently qualified, and he wished he could spend time observing and coaching him this first time, but too many higher priority jobs demanded his time. Since Wes helped write the procedure that Kelly would use, he felt confident nothing would go wrong.

With the circuit breakers out of service, it was an opportune time to overhaul the pump's sealing mechanisms. Two mechanics on the floor were waiting for him to rack out the breaker to the pump motor. Or perhaps they weren't waiting. They may have started work already, since the breaker was already open. Even though there was a lot of pressure to get things done on schedule, Kelly had to wait over an hour to get permission from the control room before proceeding with the task that really didn't need permission—it seemed like a dumb rule. In the back of his mind, he wanted to complete this job before break time. He knew the policy that if you did not take your break on time, you did not get one later. Kelly needed to act quickly—and besides, the break was scheduled to start in five minutes.

Even though the circuit breaker was already open, the procedure called for the "local trip" (open) push button to be depressed to ensure the breaker was indeed open. This was a normal safety practice to prevent personnel injury from inadvertently racking out a circuit breaker under load. Kelly was confused because the illustration in the procedure depicted push buttons different from this particular circuit breaker.

Kelly scowled at the two unlabeled push buttons in front of him. More than six months ago, he vaguely remembered talking about this breaker during classroom training, and he never got the chance to actually operate one during his On-the-Job-Training (OJT) period.

Long ago, repeated use had rubbed off any markings on the push buttons. What a stupid situation, he thought. Kelly was angry and frustrated, wanting to do a good job—to prove himself to the other electricians. One button opens the breaker to interrupt power to the pump motor. The other closes it. He looked around for a nearby telephone to call Wes, but found none. He had to choose. Perhaps there is some logic to which one is on top. Kelly figured there must be. He guessed top for "open" and pressed the button.

In that instant, the circuit closed with a disheartening loud clunk, and 4,160 volts of power surged through the line to the motor. Kelly suddenly realized his mistake and immediately depressed the other button, hoping it would trip open the breaker. It did. Fortunately, the mechanics overhauling the pump had just left the area to go on their break, and no one was hurt when the pump started. Kelly learned a valuable lesson that evening, not proceeding in the face of uncertainty. One simple action—one unwise assumption—can kill.

It would be easy to simply point the finger at Kelly, indicating what he should have or could have done to avoid such a serious near-miss tragedy. Yes, he chose to push "a" button when he was uncertain about it. But, really, under the circumstances, what else would you expect Kelly to do?

RISK-BASED THINKING—four cornerstone habits of thought

As reflected in the quotes at the beginning of this chapter, Dr. Erik Hollnagel declares that resilient performance involves the ability to adjust, while Sidney Dekker asserts that it is people who create safety. This section describes how people can adjust their performance in response to disturbances, changes, and even opportunities to not only reduce the number of things that go wrong (error and harm), but also increase the number of things that go right (CRITICAL STEPS). The more likely that something goes right, the less likely that it will go wrong.[7] The approach described in the following pages involves a way of thinking—a way of doing work, a way of managing, a way of doing business.

A first principle is a basic, foundational assumption that cannot be reasoned from any other proposition—all other conclusions and practices in a particular domain are based on one central idea. In order to sustain order in any system, if it is to minimize the occurrence of harm, it must avoid losing control of intrinsic operational hazards and minimize the resulting injury, damage, or loss, if it does. Dr. Erik Hollnagel, among other notable researchers in resilience engineering, identified four cornerstone organizational qualities important for its success and safety.[8] The exercise of control and the minimization of harm are undertaken by creating risk-relevant operational knowledge through anticipation, monitoring, responding, and learning, which collectively I refer to as RISK-BASED THINKING. If you and your organization are not using these four "habits of thought" consistently in the management of the **Hu** risk in your operations, you will likely not be successful in the long run. As a fundamental first principle of operational **Hu**— anyone can apply these habits of thought, anywhere, for any activity, any time. RISK-BASED THINKING applies to all human functions at any organizational level: plant operations, management practices, construction, maintenance, engineering design, and all the way up the corporate ladder to the boardroom.

RISK-BASED THINKING offers a means of systematically considering how to make changes that preserve safety and continue to operate. The four cornerstones of RISK-BASED THINKING include the following mental practices or habits of thought:[9]

1   *Anticipate*—know what to expect
2   *Monitor*—know what to pay attention to
3   *Respond*—know what to do
4   *Learn*—know what has happened, what is happening, and what to change.

Illustrated in Figure 4.1, these four habits of thought form the core of RISK-BASED THINKING. RISK-BASED THINKING is slow thinking, encouraging people to reflect—to know—on their work proactively, deliberately, and logically—not in a rush—to make things go right instead of mindlessly letting things happen to them.[10]

Using a sports-related principle, you are "playing to win," instead of "playing not to lose." The traditional view of safety tends to promote a play-not-to-lose mentality—avoiding harm. Although still a valid strategy, it focuses almost

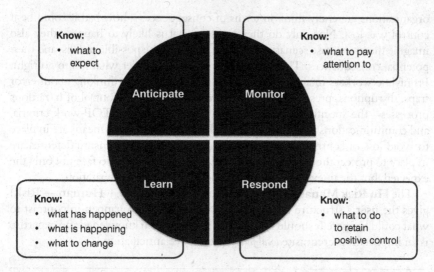

Figure 4.1 The four cornerstone habits of thought associated with RISK-BASED THINKING. RISK-BASED THINKING emphasizes knowing the facts relevant to the work at hand—**assets**, **hazards**, their **pathways**, and respective human **touchpoints**

exclusively on not doing unsafe acts or eliminating/avoiding specific hazardous conditions. RISK-BASED THINKING, on the other hand, promotes a play-to-win mentality—identifying what absolutely has to go right—do the right thing to the right thing the right way. Do what needs to be done to exercise positive control and protect **assets**. This is another way of saying safety is what you do. As introduced in Chapter 2, the **Hu Risk Management Model** offers the user a structure for thinking about **assets** and their **hazards** and what to manage—**pathways** and **touchpoints**.

RISK-BASED THINKING enables foresight and flexibility—even when surprised. Conceivably, it could be said that, *although you cannot prepare for every adverse scenario, you can be ready for anything*. When RISK-BASED THINKING is integrated into the DNA of an organization's way of doing business, people could be ready for any surprise. Let's look at each cornerstone in more detail to know what they look like in practice.

Anticipate—know what to expect

Anticipate means to foresee the implications of the task at hand—what is to be accomplished and what is to be avoided. The accomplishment specifies what's different after work—the expected changes in state **assets** will undergo during operational processes to achieve a desired work output. It's important to understand what to avoid as well—the particular harm that could occur if one loses control. First, front-line workers must think intentionally and specifically about what could go wrong in light of what is to be accomplished. Accomplishments point to **assets** and their related transfers of energy, movements of mass, and/or transmissions of information necessary to add value. People in high-reliability organizations generally think in terms of consequences—how bad it could be if control was lost. Not only do they foresee what is likely to happen, they also imagine the worst-case scenario—looking ahead for the possible pitfalls, and their potential consequences. Then, front-line workers consider what has to go right. Front-line workers imagine how they could lose control—thinking about error traps, disruptions, pressures, and how to maintain positive control of hazardous processes—the specific controls, contingencies, resources, STOP-work criteria, and communications to ensure success. This then ensures that means are in place to avoid loss of control—to exercise positive control and to ensure defenses are in place to protect the **assets** from harm. People are prepared to face not only the expected, but the unexpected as well. Anticipation leads to preparation.

The **Hu Risk Management Model (Asset + Hazard + Human → Risk)** gives the user a structure for thinking about what has to happen in contrast to what could happen. It should be apparent that technical knowledge and expertise is an important prerequisite *local factor* to effective anticipation.

Warning! The problem with the future is that more things might happen than will happen. However, front-line workers should focus on situations and consequences that pose the greatest risks to **assets**.

What does anticipation look like? Below I offer a few examples of how top performing individuals and organizations develop foresight of what could go wrong and what must go right, given the work at hand:

Front-line workers:

- Conduct pre-job briefings, where people talk about intended work accomplishments. People ask, "What are we trying to accomplish?" They talk about the desired work outputs to be achieved and the criteria for success.
- Identify the important **assets** in the work—what to protect during work.
- Acknowledge specific operational hazards for the work at hand, and the means to control and contain them—the **pathways** to harm, where **assets** are exposed to **hazards**.
- Pinpoint procedure steps and other human actions (**touchpoints**) that could trigger transfers of energy, movements of mass, or transmissions of information that alter the state of **assets**. Identify points of no return, i.e., CRITICAL STEPS.
- Ask questions, such as, "What if?" expressing doubts and differing opinions—they challenge assumptions.
- Visualize how control could be lost, given the context of the work at hand, and discuss what to do to retain positive control. Talk about possible consequences, asking, "What would happen if we lose control at a specific step in the procedure?", "What is the worst that could happen?", or "How bad could it be?"

- Create contingencies (what to do, if...) and STOP-work criteria (stop rules) to stop transfers of energy, movements of mass, or transmissions of information before **assets** suffer harm.
- Prepare for exercising positive control, but also for the worst-case outcomes.
- Review **events** and errors from previous jobs and talk about how to avoid similar occurrences.

Line managers:

- Define future operational threats and communicate them to front-line personnel in a timely fashion.
- Identify high-hazard, operational risks that require greater focus, control, and standby resources.
- Ensure training adequately focuses on the intrinsic hazards associated with the technology and emphasize RISK-BASED THINKING in skills training.

These questions help people think in terms of consequences, rather than likelihoods. We tend to underestimate likelihoods of bad things happening because we overestimate our ability to control a situation. Getting people to think about consequences is an important success factor for anticipation. Contingencies are identified—as well as gathering necessary resources needed to activate those contingencies. Once people begin thinking this way, it is natural to segue to think about what must go right—what to pay attention to.

Monitor—know what to pay attention to

Knowing what to pay attention to promotes recognition and early response to situations that could result in harm. Sound technical expertise, time, and effective anticipation help establish this knowledge. Front-line workers more readily recognize the creation of **pathways** for intrinsic process **hazards**—impending transfers of energy, movements of mass, and transmissions of information. Also, front-line workers more readily recognize the CRITICAL STEPS—**touchpoints** that absolutely have to go right the first time, every time, whether planned or unplanned. All persons engaged in the work monitor changes in critical parameters—not only for safety but also for quality. Recall from Chapter 2 that for every **asset** there is a "safe operating envelop" (SOE). Every **asset** has a set of critical parameters that define possibly multiple boundaries within which an **asset's** safety is preserved.

> *Note.* It should be apparent that monitoring is a form of real-time learning for front-line personnel. Monitoring optimizes situation awareness.

Critical parameters provide important information about changes in the state of **assets** (product, equipment, etc.), helping operators and craftsmen know where the **asset** is in relation to its safety limits. For example, for an aircraft, altitude, airspeed, fuel loading, and oil pressure are just a few critical parameters important for its safe operation. Such focused monitoring enables early response to an off-normal situation that either gets the task back on track, or helps to minimize the harm done.

Front-line workers must know and attend to an **asset's** critical parameters during particular phases of production processes—where changes in the state of product, property, or people occur; where they touch things (do work). Changes in state occur after a **pathway** is opened. Some **touchpoints**, referred to as CRITICAL STEPS, are more important than others because of the energies involved, or the volume of mass moved, or the sensitivity of the information transmitted. These **touchpoints** have to go right—the first time, every time— or else, harm could ensue. Readiness and attention must be piqued at these special **touchpoints**, which are highlighted by specific critical parameters. As one of the building blocks of H&OP and its importance to safety, an entire chapter is devoted to the topic of CRITICAL STEPS (see Chapter 5).

"Long-term survival in the wilderness depends on having the right attitude."
—Scott McMillion
Author: *The Mark of the Grizzly* (1998)

Monitoring is enhanced by an ongoing mindfulness of the safety of **assets**, which I refer to as "chronic uneasiness." Chronic uneasiness is marked by a preoccupation with harm to assets and the formation of **pathways**. (A more detailed discussion of chronic uneasiness occurs later in this chapter.) The following practices are known to improve the effectiveness of monitoring.

Front-line workers:

- Recognize the creation of **pathways** between **assets** and **hazards**. They are alert for impending transfers of energy, movements of mass, or transmissions of information that may or may not be denoted in the procedure.
- Pay close attention to the occurrence of CRITICAL STEPS and their related Risk-Important Actions.
- Check on the presence of workplace conditions that could adversely influence their control of the work—error traps, adjusting as needed to minimize their impact on their performance.
- Verify the integrity of defenses (controls, barriers, and safeguards) necessary to protect **assets** during work.
- During work, concentrate on **assets'** critical parameters most indicative of the assets' safety.

Line managers:

- Conduct in-field observations, providing and receiving feedback (see Chapter 7), including training.

- Develop and regularly review leading indicators for safety (especially those that increase as safety is enhanced).
- Monitor key indicators (lagging and leading) of plant's technical health.

While knowing what to pay attention to is an important prerequisite to creating safe outcomes, knowing what to do preserves safety. Unless front-line workers are able to respond—adjust—to workplace changes, disturbances, and surprises, the company will not long survive.

Respond—know what to do

This practice is the most important of the four. Prolonged inability to respond to threats, challenges, opportunities, or responding wrongly, could put people's health and well-being or even the organization's existence in peril. Safety, quality, reliability, and even profitability is sustained by an organization's ability to recognize, adapt to, and absorb threats, disruptions, and disturbances—as well as human error.[11] Again, this strongly suggests that safety is what you do.

"Human operators must remain in control of their machinery at all times. Any time the machinery operates without the knowledge, understanding, and assent of its human controllers, the machine is out of control."

—H.C. (Hop) Howlett
Author: *Industrial Operator's Handbook* (1995)

As a result of mindful monitoring, front-line workers take action to exercise "positive control" at CRITICAL STEPS, adjusting their actions as necessary to protect key **assets**. Positive control is defined as "What is intended to happen, is what happens, and that is all that happens." Front-line workers respond properly to make sure the right thing happens to the right thing—the right way at the right time—regardless of what procedures say, doing what is necessary, with the resources at hand, to make the situation safe for **assets**. Plant operators, surgeons, air traffic controllers, and pilots know this concept well. I'm not referring to positive control from a scientific, experimental perspective. I am referring to the positive control of operations in the workplace—positive control of transfers of energy, movements of mass, and transmissions of information. For workers to exercise positive control, two primary pre-conditions must exist. Without either, there is no control. Front-line workers must have both:

- the means to know the true state of **assets**, **hazards**, and related equipment (e.g., critical parameters);
- the capability and resources to physically manipulate or interact with **assets** and **hazards** so as to alter an **asset's** state, with the desired precision, including technical expertise (if not their own, they know how to get it), tools, time, material, etc.

So, what does responding look like? Responding involves one or more of the following: (1) eliminate, (2) prevent, (3) catch, (4) detect, or (5) mitigate. Once a **pathway** is either imminent or present, the performer may (1) choose to eliminate the task (not do it), (2) take actions to avoid losing control (error) during the task, (3) catch the occurrence of an error in the act, (4) detect the results of an error (defect or unusual condition) after the act, or (5) take measures to minimize the damage, injury, or loss if none of the previous tactics worked. The foregoing sequence is sometimes referred to as the "hierarchy of controls." The following list provides examples of behaviors and actions that help sustain positive control of transfers of energy, movements of mass, or transmissions of information and protection of assets during work.

Front-line workers:

- Before starting work, establish or verify "necessary" conditions exist to accomplish the work as planned (such as verifying initial conditions, availability of procedures and operating aids, staging of necessary tools and materials to get the work done, including placing additional safety equipment in standby for higher-risk situations or greater uncertainty (just in case).
- Compensate for missing items discovered during work, such as insufficient parts and tools or other resources necessary to complete the work.
- During work, take actions to avoid foreseeable problems or consequences, if continuance with current plan would lead to trouble. For unexpected situations (surprises) encountered during high-risk work, then stop the work in progress and obtain technical assistance. Stopping work involves suspending all transfers of energy, movements of mass, and transmissions of information related to the technical production process.
- Apply specific **Hu** tools, such as procedures use and adherence, self-checking, and peer-checking, before performing important human actions (see Appendix 3).
- Prove a work situation is safe for **assets** before initiating transfers energy, movements of mass, or transmissions of information (e.g., verify the outcomes of previously performed RIAs).
- Implement contingencies (what to do if...) and STOP-work criteria (rules to stop all transfers of energy, movements of mass, and transmissions of information) when the safety of assets is threatened. Slow down the work tempo as needed to retain positive control.
- Make conservative decisions, when production and safety goals conflict.
- Take timely actions to prevent **assets** from exceeding their respective safety limits. Otherwise, take necessary actions to minimize harm to **assets** when critical parameters for safety have been exceeded.
- *Never proceed in the face of uncertainty!*

Line managers:

- Arrange for extra safety equipment and resources for high-risk operations. Perform management oversight when needed.

- Execute emergency organizations as needed in response changing operational conditions, ensuring people with requisite expertise are engaged in operational decision-making.
- Regularly verify emergency equipment and resources are operable and available. Verify maintenance is conducted on technical structures, systems, and components as well as safety equipment.
- Avoid living with problems, correcting problems as soon as practicable after discovery.
- Allow for flexible response to off-normal situations for the purposes of safety, when procedures and expectations are unclear.

Before beginning work, supervisors and workers develop contingencies (what to do if...), "hold points" (for important checks before proceeding), and abort criteria (when to stop) so as to respond appropriately when instructions are unclear, things don't go as planned, or they are otherwise surprised. These discussions are forms of anticipation. Proactively, they identify *ways* to exert positive control on the right thing, at the right time, in the right way—consistent with the asset's need for safety.

But, if you are unsure you can ascertain the safety of an **asset**, STOP and get help, if at all practicable. When time is available, stopping any transfers of energy, movements of mass, or transmissions of information is preferable, especially when uncertainty arises at CRITICAL STEPS. Error rates associated with uncertainty are usually high. Technical expertise is needed when uncertainty arises. Successful adjustments to uncertainty—the ability to cope effectively with an immediate threat—depend greatly on people's technical knowledge of the system (expertise) and accurate situation awareness. However, I believe anticipating, monitoring, and responding well depends on learning.

Learn—know what has happened, what is happening, and what to change

Workplace learning occurs when a person does something in response to new information—when behavior changes. Just as risk is dynamic, learning must necessarily be dynamic—continuous—occurring not only before, but also during, and even after work. For every high-risk task in the workplace, front-line workers consider previous work experiences (the past). Knowing what went wrong previously (known as operating experience), whether locally or within the industry, provides insight into what to avoid "the next go around." Also, it is worthwhile to understand how the job went right last time in spite of the conditions that were encountered.

Systems-thinking managers understand that there are no perfect procedures or work plans, and that risks change dynamically during work. Therefore, managers promote a chronic uneasiness among the workforce to enable them to readily recognize surprise **pathways** and the appearance of CRITICAL STEPS—a form of real-time learning. Awareness of current risk conditions and activities during the task at hand (the present) enables front-line workers to either halt the operation that could cause harm or make appropriate adjustments before proceeding so as to protect important **assets** from harm.

Opportunities for learning happen also after the work is completed. Because *work-as-done* almost always differs from *work-as-imagined*, front-line workers are encouraged to report significant differences between them, even when nothing bad happens. These differences, whether for good or bad, offer managers keen insight into the effectiveness of their management systems. Supervisors and the principal front-line workers capture such learning during brief meetings referred to as post-job reviews or after-action reviews (see Appendix 3). Such information can be tracked to resolution using Corrective Action / Preventive Action (CA/PA) process.

Finally, learning demands humility. Without humility, no one learns. Without "intellectual humility," you are unable to learn.[12] Intellectual humility acknowledges your own fallibility and that others may know something about the operation that you don't know. If you have no doubts about the integrity of your system and its defenses you tend not to listen and stop asking questions. Humility does not come easy to any leader.

Beside the classical classroom training venue, learning occurs several ways in multiple locales in the workplace. Here are a few practical examples of what learning looks like operationally:

Front-line workers:

- Before starting work, take time to discuss lessons learned from previous challenges, **events**, and successes for the task at hand. People pinpoint what will be done differently to avoid previous errors and **events** and to ensure success.
- Maintain real-time situation awareness of critical parameters—continuously updating one's mental model of the technical process.
- Challenge assumptions and misunderstandings, correcting at-risk practices of co-workers, supervisors, and even managers. You *are* your brother's keeper!
- Call periodic timeouts from the work during natural pauses, especially before commencing critical phases, to discuss with co-workers current work status, future actions, and potential pitfalls to avoid.
- Defer to those with requisite technical expertise for important operational decisions.
- After completing work, take time to discuss surprises—differences between *work-as-done* and *work-as-imagined* (post-job reviews), close calls, and other adjustments needed to achieve desired accomplishments; follow up by submitting a report of significant issues experienced during a task.

Line managers:

- Regularly observe pre-job briefings, work in progress in the field, and post-job reviews, engaging front-line workers to better understand how they are successful, despite prevailing workplace conditions.

- Practice high-consequence scenarios during simulator/laboratory training and emergency drills.
- Ensure front-line personnel take advantage of operating experience.
- Thank and reinforce people for safe work practices. Coach and correct those who exhibit unacceptable or at-risk practices. Be specific—avoid generalities.
- Schedule time for front-line workers to participate in refresher training— to keep them abreast of procedure and equipment changes, to refresh fundamental knowledge, and to study relevant operating experience reports.
- Actively encourage a willingness to reveal personal errors, shortcuts taken, and near-misses—reinforcing people and avoiding punishment in any form when people share such information. Assure subordinates that they will not be punished for isolated, one-off "honest mistakes."[13] The dignity of persons is upheld and valued.
- Are accountable for using a CA/PA process.
- Promptly respond to reports from front-line workers.
- Follow through with learning from reports, benchmarking, self-assessments, and adverse trends using a corrective action / preventive action tracking system, which managers are held accountable to.
- Adopt a systems view versus a person view during analysis of **events**.

In summary, RISK-BASED THINKING enhances the real-time risk assessment during operations—it's an operating philosophy that everyone at all levels of the organization can apply regularly without referring to a procedure. These four habits of thought form the basis of resilience—the flexibility to adapt as needed to protect **assets** from harm when uncertainty arises. Integrating RISK-BASED THINKING into line operations helps line managers, supervisors, and front-line workers systematically think about, plan for, and respond to critical situations— whether expected or unexpected. RISK-BASED THINKING combined with the **Hu Risk Management Model (Asset + Hazard + Human → Risk)** helps people recognize **pathways** and CRITICAL STEPS, including those not denoted in a procedure. Any human function—whether operations, maintenance, engineering, clerical work, and even top-level management—can benefit from using this first principle. Table 4.1 summarizes some of the key attributes of RISK-BASED THINKING.

Chronic uneasiness

As the Engineering Officer of the Watch (EOOW) responsible for the operation of the nuclear propulsion system and related engineering spaces of a nuclear submarine, I would walk through the plant once during each six-hour shift to inspect the plant's material condition and housekeeping, check on critical parameters of important equipment, and assess the knowledge and alertness of the enlisted operators. I looked forward to this break from the cramped confines of the submarine's "Maneuvering Room." The "gemba walk" took me alongside process equipment, propulsion systems, and a nuclear reactor operating at high

Table 4.1 Summary of the key attributes of RISK-BASED THINKING

Anticipate	Monitor	Respond	Learn

• Review procedures and other guiding documents, including emergency procedures before starting work
• Know desired accomplishments (changes in state of assets) and criteria for success (business objectives)
• Pinpoint **assets** and intrinsic **hazards** for work at hand
• Know when **pathways** and related **touchpoints** will occur in work
• Know what to avoid (harm due to loss of control of **hazards** during work)
• Ask "What if…?"
• Develop contingencies and STOP-work criteria
	• Be mindful of and recognize **pathways** that poise the production system to:
	  ◦ transfer energy
	  ◦ move mass
	  ◦ transmit information
	• Recognize **touchpoints**, especially CRITICAL STEPS
	• Verify integrity of defenses
	• Check and validate **asset's** critical parameters regularly during work (safe operating envelop)
	• Establish management oversight for high-risk operations
		• Prove work is safe before proceeding. Never proceed in the face of uncertainty!
		• Stage additional resources just in case to respond to surprise situations
		• Follow approved procedures and expectations unless they will be harmful if followed as written
		• Exercise positive control at CRITICAL STEPS during work
		• Make conservative decisions
		• Use **Hu** tools as needed
		• Apply contingencies, STOP-work criteria, or emergency procedures when needed
		• Take timely actions to protect assets from harm with whatever means
			• Be humble
			• Recall operating experiences of similar jobs and discuss how to avoid relevant mis-steps for tasks at hand
			• Call timeouts before CRITICAL STEPS, stopping work momentarily to verify everyone understands the current risk situation
			• Defer to those with expertise for operational decisions
			• Managers engage front-line workers during field observations
			• Report significant differences between *work-as-done* and *work-as-imagined*
			• Hold managers accountable for CA/PAs
			• Managers adopt a systems view of performance.

temperatures, high pressures, high voltages, high RPM, and in the presence of nuclear radiation. However, there was one place in the propulsion spaces that always gave me the jitters. I was always fearful to stand or even walk near the 4500/3000 psi high-pressure air-reducing station. A pin-hole leak at such high pressures would pierce your flesh faster than a hot knife through butter. You could say I was uneasy.

> "I am always scared. Imagination and fear are among the best engineering tools for preventing tragedy."
>
> —Henry Petroski
> Author: *To Engineer is Human* (1985)

A trait of high-reliability organizations (HRO) related to a healthy chronic uneasiness is that its people are "preoccupied with failure"[14]—one of five key tenets described in *Managing the Unexpected* by Karl Weick and Kathleen Sutcliffe.[15] Although this principle was originally oriented toward learning from small failures, the attitude advocated by Weick and Sutcliffe is just as applicable to anticipating and avoiding harm in real time.

A true professional's mind-set for every job, regardless of how simple or routine, is to perform the work error free. However, it is not uncommon that performers in high-hazard industries are most at risk when they have gained enough experience to become comfortable with the risks they encounter regularly.[16] Remember the RCA Building ironworkers? After a while on the job, they began to believe they worked in a safe environment (because nothing bad has happened) rather than in an intrinsically dangerous one. True professionals, regardless of competency, proficiency, and experience, possess an ongoing wariness of the **assets** and **hazards** in their work—chronic uneasiness. They understand implicitly that they work amidst multiple **hazards** in an unforgiving environment that is subject to unexpected occurrences and unfamiliar or hidden conditions. Even though they expect success, they are mindful of any action that involves a transfer of energy, movement of mass, or transmission of information. They exercise positive control of those actions to avoid critical mistakes that could trigger a loss of control of hazards.

> "If you don't have doubts, you haven't been paying attention."
>
> —unknown

But vigilance is difficult to sustain, and these high-risk situations are not always highlighted in procedures. Too often people go about their day-to-day business unaware of **hazards** around them. If "nothing has ever gone wrong," they will likely think that "nothing will go wrong today." Routine success desensitizes people to the threat of harm, while all their focus is directed toward achieving ever-present production objectives. A chronic uneasiness guards against this bias commonly referred to as complacency. I think John Krakauer, in his book *Into Thin Air*, summarizes this unhealthy attitude most succinctly in the following excerpt:[17]

> ... the sort of individual who is programmed to ignore personal distress and keep pushing for the top is frequently programmed to disregard signs of grave and imminent danger as well. This forms the nub of a dilemma that every Everest climber eventually comes up against: in order to succeed you must be exceedingly driven, but if you're too driven you're likely to die. Above 26,000 feet, moreover, the line between appropriate zeal and reckless summit fever becomes grievously thin. Thus, the slopes of Everest are littered with corpses.

Summit fever can blind you to the occurrence of **pathways** and CRITICAL STEPS. The following attitudes characterize the two extremes of people with and without a chronic unease:

• People *without* a chronic unease: "I've done this job every day for the last month, and nothing ever changes. I don't need to worry about it; I've got this."
• People *with* a chronic unease: "I've done this job every day for the last month, but something could be different. I'm working with some pretty nasty stuff. What could go wrong? Given the circumstances, what am I going to do to maintain control of my work?"

People in the first category usually are highly skilled in their work, proficient, and possess extensive experience. Or, they may have been extremely lucky! Top performers, on the other hand, are continuously mindful of two realities during their work. First, they have a lingering anxiety with transfers of energy, movements of mass, or transmissions of information that occur during their work, especially those associated with key **assets**. They continually update their mental picture of **pathways** of potential harm to key **assets**, continually learning, in real time. Second, they're humble—they understand and acknowledge their own fallibility, blind spots, and their capacity to err. Regardless of experience, training, qualifications, education, and proficiency with a job, they know that making a mistake is always possible. They know that not everything is always as it seems and that they don't know all there is to know. Land mines, error traps, insufficient resources, sudden changes in tempo, and unanticipated situations lurk in the workplace that could hurt someone or something if they let their guard down—even just for a moment.

> "The thing that I really like about traveling or hiking in country where you know there are grizzly bears is that your hearing improves, your eyesight improves, your sense of smell improves. You're paying attention a lot more. You're a lot more alive if you are paying attention to bears. And, if you're not paying attention, you can wind up a lot more dead."
>
> —Tom Murphy
> Wildlife photographer

Front-line workers can never really relax while handling important **assets** and their intrinsic **hazards**.[18] Nothing can be taken for granted. As Gene Kranz of Apollo 13 fame (launched in the spring of 1970) is known to have said, "Failure is not an option." People with a chronic uneasiness insist on facts— they avoid assumptions and ambiguities when operational decisions pose a risk to **assets**, or place positive control of **hazards** in jeopardy. A chronic uneasiness encourages people to stand ready to revise their risk assessments—as new, confirming or contradictory information surfaces.[19]

Personal experience: following GPS instructions

Following procedures, checklists, and expectations is a lot like following the computer-generated instructions of a GPS (global positioning system) device. My grandson was scheduled to be the starting pitcher in a baseball game late one Saturday afternoon. The contest was at a park north of Atlanta that I had visited once before, but I was still hazy on how to get there. So, as many of us do, I entered the name of the park into the search field of my GPS app on my "smart" phone, found the place, and started navigation from my current position. GPS directions don't always correspond to the printed road signs, and the "turn in X feet" announcements do not always correspond to the correct street. Or it may direct you down a one-way street, or into a construction site. I knew roughly where the park was, so I wasn't too concerned.

Things went along smoothly until I heard an instruction to continue on a particular road for several miles, which in my gut I knew was wrong. From a previous visit, I recollected turning much earlier. I went with my gut and turned. I recognized the route and continued on. Meanwhile, the GPS is spouting out instructions to take the next turn, repeating new instructions each time I passed a designated turn. Finally, I arrived at my destination, and the GPS proudly announced, "Arrived." The route I chose was quicker and shorter than the suggested GPS route. Not sure what the GPS unit was "thinking." I continue to use my GPS, but I treat the directions with skepticism, particularly in an area that I don't know well. GPS is very helpful, but not perfect.

As S.I. Hayakawa[20] put it, "The map is not the territory." A map is a frozen-in-time likeness of the territory (reality). Reality is dynamic, not static. Factors to consider in following a GPS application:

- The software designer assumed certain conditions at the time of the app's development.
- GPS is not real-time.
- GPS cannot detect current conditions.
- The user has to adapt to actual situations.

Similarly, operators should follow their procedures and checklists with a measure of caution and skepticism. One commercial nuclear electric utility in the Midwestern United States adopted the following catch phrase to promote a chronic uneasiness toward work: "Every day, every job, error free—treat nothing as routine." Now that's wisdom to live by.

Following technical procedures rigorously is most important to safety. However, too often people focus inordinately on compliance or avoiding mistakes, instead of what's happening operationally. Do not let an operator or technician follow a procedure or checklist mindlessly as if it is unquestionably correct in all respects. *Just as the map is not the territory, the procedure is not the system.* Encourage your front-line workers to follow procedures cautiously as they monitor and understand what is really happening with the work—with **assets** and **hazards** for every written step.

In summary, a chronic uneasiness is best described as an ongoing mindfulness about the work at hand exhibited by the following attitudes:

- a preoccupation with impending or current transfers of energy, movements of mass, or transmissions of information that could harm an **asset** if performed out of control;
- an awareness of the presence of hidden **hazards** in the workplace, such as land mines, faulty defenses, and error traps;
- a preoccupation with changes to critical parameters of an **asset**;
- a humble acknowledgement of one's fallibility—the capacity to err;
- a reluctance to consider any job or situation as "routine."

The front-line worker—hazard or hero?

Humans are known to hurt themselves and others, to break things, and to lose things, among other lifelong blunders—it's in our genes. *Errors happen!* There are no such things as perfect procedures, perfect knowledge, or perfect defenses—people, workplaces, operations, and intrinsic risks are dynamic, changing day to day, often moment by moment. **Events** *happen!* When errors trigger **events**, too often people are blamed for their "lack of judgment" or "carelessness." However, people, though fallible, do much more than simply follow rules—they think and adapt.[21] Despite their fallibility, it's only front-line workers—the specialists in their work—who can see these variations in risk and are capable of adapting correspondingly. Yes, they are indeed a **hazard** but also the hero.[22]

For hundreds of years, workers have been perceived as instruments or tools used to accomplish someone else's goals. In days past, most work was physical, observable in nature. A worker's job was simply to do what he/she was told to do, no more, no less. That translated into compliance.

Today, the preponderance of work is more information-based. Most work is a combination of physical and mental activity. Much of the activity goes on in the heads of those who do the work. Given the variable nature of work and the workplace, workers must be able to adjust their practices to meet the rise and fall of risks where they are. Workers are not tools. If treated as such—as followers, people will become passive, doing only what they are told to do—an unsafe climate anywhere. As some may say, "The best way to stay out of trouble is to follow procedures without question, not do anything else, or make decisions."[23] Alternatively, I strongly suggest that front-line workers be perceived and treated more as "agents" of the organization—representatives of the organization—enabled to make safe adjustments consistent with the values and priorities of the organization's leadership. Dr. Fred Nichols, Distance Consulting, LLC, says the following about how employees should be perceived by their managers:[24]

> Instead of simply carrying out prefigured routines, employees must now configure their responses to the circumstances at hand and these typically vary from situation to situation. Interactions are now between people and information and people. In short, employees, not industrial engineers, must now figure out what to do and how to get it done. What is now wanted from employees are their contributions, not their compliance. To contribute, employees must exercise discretion and to do that they must be granted an appropriate degree of autonomy. Today's employees are better viewed as agents, acting on their employer's behalf and in their mutual best interests, instead of as mere instruments of managerial will.

People like to accomplish goals, make things better, be effective, and be efficient at the same time. A work context fraught with unclear procedures, insufficient tools and resources, poor working conditions, improperly configured process systems, and worn out equipment, tends to lead to adjustments on the job—most for good but some for ill—all products of the way your system works.[25] Adjustments in the workplace often look like errors in the form of faster vs. slower, out-of-sequence actions, earlier vs. later, different direction, etc., except that legitimate adjustments are intentionally oriented in accomplishing one or more goals. In his book, *Safety-I and Safety-II*, Dr. Erik Hollnagel describes three reasons (general work situations) why people adjust their performance.[26]

- *Maintain/create conditions for use in the future.* These are activities taken (based on previous experience) to establish or preserve conditions necessary to accomplish the work, or conditions important to comfort, coordination, timeliness, access to and availability of resources, safety, etc. For example, working a little faster to create time later, adding individuals for specific tasks to avoid overloading one individual, arranging engineers to stand by their telephones to respond to technical questions; adding temporary lighting to the workplace, etc. Mostly, these activities ensure sufficient time is allotted.
- *Compensate for something missing.* People respond to the discovery of some unacceptable conditions that occur during a job that impede accomplishment of work, such as workarounds, broken tools, inoperative indicators of critical parameters, unavailable parts, insufficient time, missing technical data or information, malfunctioning equipment, weather, etc. Such situations tempt people to make trade-offs or take shortcuts.
- *Avoid future problems.* People take actions to avoid foreseeable, negative consequences to themself, to the work group, or to the organization, if continuing unchanged would surely lead to problems. For example, changing a schedule; adding ventilation to avoid heat-related issues; avoiding interruptions with a "do not disturb" sign, etc.

Don't think adjustments are inherently bad. In reality, most adjustments by front-line workers are necessary to maintain control. It's a good thing when workers make adjustments for safety, whether or not the adjustments follow procedures and adhere to management expectations. Information about one's work is never fully complete, and the time available is rarely adequate. Therefore, front-line workers need the flexibility to adjust—to adapt—when the safety of **assets** is in jeopardy. Managers should give front-line workers the authority to always protect **assets**—to deviate from procedures or stop operations when they are unsure—to act as the organization's agents to do what is prudent to avoid injury, damage, or loss, and to get help. This is policy-level guidance that should clarify the boundaries and situations associated with such decisions—such guidance must be explicit and trained on before giving people freedom to deviate from procedure—to act responsibly as the organization's agents in the workplace.

Technical expertise—the bedrock of RISK-BASED THINKING

An **event** occurs when an **asset** suffers harm—the boundaries of what is safe for an **asset** is exceeded. If front-line workers are to avoid harm to **assets** during operations—when procedures and processes provide insufficient guidance—they must possess sufficient technical knowledge and understanding of the safety boundaries for all the **assets** they work with on the job.

I served seven years on active duty in the United States submarine service, during which time, I experienced firsthand the emphasis on technical knowledge and skill in the Naval Nuclear Power Program. The U.S. Navy has launched more than 200 nuclear-powered ships, using 30 different power plant designs, with over 500 reactor cores brought into operation since the start of the nuclear power propulsion program in 1948. As of 2006, those ships logged more than 5,700 reactor-years of operation and traveled well over 134 million miles. Since the launch of the U.S.S. Nautilus (SSN 571) in 1950, the Navy has not suffered a single, reactor-related casualty or uncontrolled release of radiation.[27] The officers and sailors who operate these nuclear propulsion systems know how these systems work and their safety limits.

Admiral Hyman G. Rickover, the revered "father of the nuclear navy" (now deceased), considered training a core value, and it continues to be so to this day. Highly knowledgeable and well-skilled operators and leaders cannot be considered a luxury where safety and resilience are concerned. There is no room for guessing. At the slightest indication of vagueness or ambiguity about safety, Admiral Rickover demanded facts. A "reluctance to simplify"[28] is another key tenet of high-reliability organizations (HRO).[29] Deviations from normal operating conditions were readily recognized and consistently reported—I remember there was an exceptionally low threshold for what counted as a problem or incident. To recognize a deviation (to see problems), Admiral Rickover demanded that his officers and sailors possess detailed technical knowledge of the submarine's reactor systems, equipment, and machinery spaces for which they were responsible. Without such knowledge, operators may know what to do, but not know why. Admiral Rickover continually emphasized training on reactor physics, fluid flow, thermodynamics, etc. as expressed by him:

> "We train our people in theory because you can never postulate every accident that might happen...[T]he only real safety you have is each operator having a theoretical and practical knowledge of the plant so he can react in any emergency."
>
> —Hyman G. Rickover[30]

Admiral Rickover believed that to manage the operation of a nuclear submarine propulsion system (or any high-risk industrial operation for that matter), you have to fundamentally understand its technology.[31] This was *not* a general overview of how things worked. Rather, it was a detailed understanding of the engineering, science, chemistry, and physics behind the processes. The requirement for technical knowledge didn't stop with the Engineering Department Officer. The technical knowledge requirements went all the way from the enlisted personnel (operators) up to the submarine's commanding officer (executive). In the U.S. Submarine Force, the higher you were on the organization chart, the better your technical knowledge was supposed to be.

Because of their expertise, your best performers likely have a *deep-rooted respect for the technology* they work with. This is more than simply knowledge. Expertise includes understanding, experience, and proficiency. Practitioners, operators, and craftsmen not only understand the safe and proper means for transfers of energy, movements of mass, or transmissions of information, they also understand when and how **pathways** between **assets** and intrinsic **hazards** are created and how that could potentially cause harm—if they lose control. Top performers continuously update and calibrate their awareness of **hazards** and their proximity to **assets**. Consequently, they more readily anticipate the worst, recognize the mistakes they dare not make, and equip themselves to respond appropriately.[32] They readily recognize when production activities could harm assets.

> "Control without competence is chaos."
>
> —L. David Marquet
> Author: *Turn the Ship Around* (2012)

Conservative decision-making

Conservative decision-making is a deliberate means of thinking and deciding, and is used to anticipate the potential effects of a workplace decision on the safety of specific **assets**, in the here and now. Conservative decisions are followed with a plan of action that preserves the integrity of **assets** and the effectiveness of their defenses despite the unyielding presence of production pressures. Conservatism is necessary to allow for possible unknown and unforeseen effects.[33] This is the essence of conservative decision-making and the long-term, reliable operations it produces.

Warning! Never proceed in the face of uncertainty! Do not assume safety is present—verify work can be accomplished safely for all **assets** associated with the work at hand. Safety is what you do. *Prove it safe!*

Profit goals, production pressures, and financial constraints are facts of life in the marketplace and always oppose safety. In the heat of the moment, when safety and production goals conflict, safety should be the default response. Executives and their line managers must take care to inculcate a worldview of risk in the workplace that avoids being blind to the consequences of one's actions.[34] People must believe that preserving the safety of **assets** for the long term is always more important than achieving near-term production goals. Otherwise, people will tend to sacrifice safety in deference to production.

To make a conservative decision one must presume that the decision-maker possesses the requisite technical knowledge and know-how to understand where an **asset's** boundaries for safety are and what to do to preserve those boundaries. (Recall the safe operating envelope.) Without proper technical expertise, the ability to make conservative decisions is suspect. With this in mind, I propose the following thought process for making a conservative decision—*GRADE*:

1 *Goal.* Clarify the desired safety state (goal) for key **assets** for the current situation, using the **assets'** critical parameters and their limits (administratively and physically) before taking any action.
2 *Reality.* Recognize the current state of key **assets** (critical parameters) and their **pathways** for harm as they arise. Demand facts. Avoid assumptions. The difference between the goal state and current reality defines the safety problem.
3 *Analysis.* Determine what has to happen to close the safety gap—to ensure the integrity of **assets** and their defenses (typically barriers). Assess the ability and time needed to restore a safe situation—buy time, if practicable. Develop options that will restore the goal safety state. Anticipate what will happen for each option. Determine the costs, benefits, and risks of each option (if time exists).
4 *Decision.* Choose and execute a course of action that will achieve the goal safety state. Decide who will do what by when. If time is scarce, adjust the response accordingly with the resources available to place the **asset** in a safe condition. Identify the key process indicator(s) of the **asset's** goal safety state (critical parameters). Consider what to do if things do not go as expected.
5 *Evaluation.* Monitor how an **asset's** critical parameters vary in response to the actions taken. Review the effectiveness of the decision in achieving the goal safety state. Perform a post-job review or after-action review after completing the work or at the end of the shift.

Notice that the depth of analysis and the process of decision-making varies depending on the seriousness of the issue and the availability of time. The logic of the process lends itself to use in rapidly changing situations just as it would for more low-tempo operational scenarios. Also, notice that conservative decision-making is fundamentally RISK-BASED THINKING applied—the four cornerstone habits of thought are present in each step of GRADE. To embed conservative decision-making into the workforce—to make it a habit, people must be trained on it and given frequent opportunities to practice using it under stress in operational settings.

Given our human limitations, unable to see and understand everything that is happening at once, it would make sense to give the front-line workers, those closest to the **assets** and intrinsically hazardous work processes, the ability to make conservative decisions to protect **assets** when the system would otherwise lead to harm. Just as it takes teamwork to suffer an **event**, it takes teamwork to preserve safety. Remember, managers are not perfect and need backup, just as front-line workers need backup. A special "culture" has to exist that gives front-line workers such flexibility, welcoming them as agents of the organization—representing the best interests of other workers and the organization.

Safety is a core value

Unless concern about the protection of **assets** is explicitly and compellingly fostered, production behaviors will naturally deter protection behaviors. Conservative decisions are more likely when most people adopt safety as a core value (from the heart) rather than treat safety as just another priority or a goal to be measured.[35] When considered a core value, safety is something they will do regardless of their priorities, resource constraints, pressures, etc. Doing things safely becomes an essential part of all operational practice, not an add-on. This is a leadership challenge.

I always chuckle whenever I see a large vehicle moving down the highway with a bumper sticker that displays the slogan, "Safety is my goal." Of course, that's never true. If it was, then the driver of that vehicle would promptly park it. It's a noble thing to promote safety in such ways, but you have to be honest with yourself. Line managers are particularly susceptible to situations where production and safety demands suddenly compete. In the urgency of the moment, managers, supervisors, and even front-line workers tend to believe safety exists since "we haven't had any **events**"—and push ahead with production activities, unknowingly sacrificing safety margins for **assets**.

Healthy, resilient organizations do not sacrifice their risk-management activities to save money in hard times; neither do they assume their operations are safe just because nothing bad has happened. H&OP and RISK-BASED THINKING do not wax and wane with the turnover of people, resources, and profitability.[36] When integrated throughout an organization, RISK-BASED THINKING stimulates mindful analysis and reflection of **assets** and their **hazards** and the occurrence of **pathways**, high-risk **touchpoints**, and how to control them.

Nancy Leveson, in her book, *Engineering a Safer World*, emphasizes the emergent nature of safety and risk in organizations:[37]

Safety is an emergent property of the system that is achieved when appropriate constraints [defenses to protect **assets**] on the behavior of the system and its components [people, processes, equipment, and **assets**] are satisfied. ... Instead of defining safety management in terms of preventing component failures, it is defined as creating a safety control structure that will enforce the behavioral safety constraints and ensure its continued effectiveness as changes and adaptations occur over time. [Information in brackets added.]

Risk is dynamic—**hazards** come and go throughout the production workday. Just as one does during his/her commute to and from work each day, front-line workers must constantly adjust their functioning to match emerging operational conditions that pose new threats to an **asset's** safety. Just as temperature, pressure, and density are properties of water, managers of top-performing companies understand that safety is a dynamic and ever-changing property of the organization. The current level of safety is an outgrowth of how the organization functions and operates—technically, organizationally, and socially—an emergent property of the system.[38]

So, how do you know that safety is present? What evidence do you rely upon to confirm or prove that the workplace is safe—for every task, job, and operation? Safety is *not* only a condition—procedures, locks, fences, passwords, personal protective equipment! Risks and dangers are not always stable, and in some industries, they are in constant flux. Doesn't it make sense that safety, too, must be responsive and dynamic? To verify the presence of safety, I encourage you to look at the behaviors people choose in response to changing risks. Safety is what people do to protect **assets** against a loss of control of intrinsic **hazards** during operations. Just as in war, friendly forces must adapt to counter the enemy's moves.

Personally, I don't like the idea of "balancing" safety and production. What does it mean to balance safety and production? If I must sacrifice a safety margin to achieve a production objective, then something is wrong in the design of the work system. If safety of **assets** cannot be assured—*always*, then the production process must be interrupted, and subsequently modified, until safety can be guaranteed. This is another reference to "conservative decision-making." Whenever there is a conflict between production and safety goals, the default decision is to assure the safety of **assets**. If you cannot operate safely, don't operate!

RISK-BASED THINKING is an operating philosophy that influences not only the way front-line workers do their job but also how managers and supervisors do theirs. Safety, when accepted as a core value, will not be sacrificed as production demands change with the marketplace.

Things you can do tomorrow

1   Is there a corporate policy on when and how front-line workers can exercise autonomy in the workplace for the purpose of safety? If yes, is it clear, and are people adequately trained on how to make conservative decisions? If no, what elements would such a policy address?

2   Periodically refresh people's memories of past significant events within the company and industry. Reinforce RISK-BASED THINKING and chronic uneasiness during the conversation. Check if operating experience is incorporated into formal training programs.

3   Discuss with your colleagues and/or direct reports what it means that RISK-BASED THINKING is a "fundamental first principle."

4   Whenever you talk about production goals, activities, and schedules, specifically address the safety implications and related controls and defenses, concurrently.

5   To reinforce RISK-BASED THINKING as an operational norm, management teams should reserve time on their production meeting agendas to discuss safety. In particular, for scheduled operations, line managers could brainstorm likely workplace scenarios that could create safety-production conflicts for front-line workers. Guidance for making conservative decisions would be made for those scenarios and then communicated to appropriate production personnel.

6   Adopt the language of H&OP in your interactions with others (see Glossary).

7   Realize that H&OP is *not* simply about preventing error, improving performance, achieving excellence, or accountability. It's a way of thinking about and managing risk. It's an operating philosophy. Discuss with your colleagues or direct reports what their operating philosophy is. Identify how it corresponds to or differs from RISK-BASED THINKING. Do you need to formalize your operating philosophy?

8   Think about how RISK-BASED THINKING and chronic uneasiness can be incorporated into your operation. Ask your management team for their opinions and input.

Notes

1   Hollnagel, E. (2016). *Resilience Engineering*. Retrieved 8 May 2017 from http://erikhollnagel.com/ideas/resilience-engineering.html.

2   Quote taken from video by Arnold, R. (2008). Human Factors and System Safety: *Dekker on Resilience*. Leading Opinion Films. Retrieved 14 November 2016 from https://www.youtube.com/watch?v=mVt9nIf9VJw.

3   Dekker, D. (2011). *Drift Into Failure: From Hunting Broken Components to Understanding Complex Systems*. Farnham: Ashgate (pp.176–178).

4   Hollnagel, E. (2014). *Safety-I and Safety-II: The Past, Present, and Future of Safety Management*. Farnham: Ashgate (pp. 44–45).

5   Ibid. (p.127).

6   Industrial circuit breakers, devices for supplying and interrupting electric power to various electrical components, are typically contained in a cubicle and connected to a heavy conductor, known as a bus (power supply). When a breaker is open, it can be rolled away from the back of the cubicle on tracks so that it disconnects from the bus; this is known as racking the breaker; usually accomplished by means of a rotary crank or lever.

7 Ljunberg, D. and Lundh, V. (2013). *Resilience Engineering within ATM—Development, adaption, and application of the Resilience Analysis Grid (RAG)*. Linköping: University of Linköping, LiU-ITN-TEK-G--013/080--SE. (p.1).

8 The concept of *resilience engineering* looks at individual performance through the lens of the organization and the system it creates so as to cope with changing risks in the workplace—the capacity to adapt. As an outcome of the way the organization is designed and managed, resilience is thought of as an emergent property of an organization as temperature is for water in response to its environment. Erik Hollnagel describes the basic qualities for resilience in the Epilogue: Resilience Engineering Precepts in Hollnagel, E., Woods, D. and Leveson, N. (Eds.) (2006). *Resilience Engineering: Concepts and Precepts*. Aldershot: Ashgate (pp.348–350).

9 Hollnagel, E. (2009). The Four Cornerstones of Resilience Engineering. In Nemeth, C., Hollnagel, E. and Dekker, S. (Eds.). *Resilience Engineering Perspectives, Volume 2: Preparation and Restoration*. Farnham: Ashgate (pp.117–133).

10 Dr. Erik Hollnagel, via personal communication, prefers to refer to the cornerstones alternatively as "opportunity-based thinking." He makes an important point that resilience engineering is all "about what we want to achieve rather than about what we want to avoid." I'll leave it up to the reader as to what you call the cornerstones—either way, they work.

11 Woods, D., Dekker, D., Cook, R., Johannesen, L. and Sarter, N. (2010). *Behind Human Error* (2nd edn). Farnham: Ashgate (p.38).

12 A concept originated by Sir John Templeton (1912–2008), an American-born British investor, fund manager, and philanthropist. Intellectual humility is a virtue, a character trait that allows the intellectually humble person to think and reason well. It is plausibly related to open-mindedness, a sense of one's own fallibility, and a healthy recognition of one's intellectual debts to others. Retrieved 3 May 2017 from https://www.templeton.org/grant/the-philosophy-and-theology-of-intellectual-humility.

13 An "honest mistake" is a colloquialism for an error committed without fraud, deceit, malice, or recklessness. Really, there are no such things as honest or dishonest mistakes. Error is always unintentional.

14 A *preoccupation with failure*, as described in *Managing the Unexpected*, involves a mindfulness of anticipation of what and how things could go wrong and a posture for learning from small failures or deviations—articulating what should not fail and how they possibly could fail.

15 Weick, K. and Sutcliffe, K. (2007). *Managing the Unexpected: Resilient Performance in an Age of Uncertainty* (2nd edn). San Francisco, CA: Jossey-Bass (p.9).

16 Rochlin, G. (1999). The Social Construction of Safety. In Misumi, J., Wilpert, B. and Miller, R. (Eds.). *Nuclear Safety: A Human Factors Perspective*. London: Taylor & Francis (p.12).

17 Krakauer, J. (1997). *Into Thin Air*. New York: Anchor (pp.185–186).

18 Hollnagel, E. and Woods, D. (2006). Epilogue: Resilience Engineering Precepts. In Hollnagel, E., Woods, D. and Leveson, N. (Eds.). *Resilience Engineering: Concepts and Precepts* (pp.355–356). Aldershot: Ashgate.

19 Woods, D., Dekker, S., Cook, R., Johannesen, L. and Sarter, N. (2010). *Behind Human Error* (2nd edn). Farnham: Ashgate (p.91).

20 Samuel Ichiye Hayakawa (1906–1992) was a Canadian-born, American academic and political figure. He was an English professor, served as president of San Francisco State University, and then as a United States Senator for California from 1977 to 1983.

21 Woods, D., Dekker, S., Cook, R., Johannesen, L. and Sarter, N. (2010). *Behind Human Error* (2nd edn). Farnham: Ashgate (p.8).

22 Reason, J. (2008). *The Human Contribution: Unsafe Acts, Accidents, and Heroic Recoveries*. Farnham: Ashgate (p.3).

23 Marquet, L. (2012). *Turn the Ship Around!* New York: Penguin (pp.38, 43–44).

24 Nichols, F. (November 2015). Employees: Instruments or Agents? [Web log post]. Retrieved November 27, 2015 from https://www.linkedin.com/pulse/employees-instruments-agents-fred-nickols-cpt?trk=hb_ntf_MEGAPHONE_ARTICLE_POST.

25 This phenomenon is commonly referred to as *emergence*—the unpredictable outcomes of the interactions and behaviors of the components and processes of a complex system—the most variable component being people.

26 Hollnagel, E. (2014). *Safety-I and Safety-II: The Past, Present, and Future of Safety Management*. Farnham: Ashgate (pp.156–159).

27 United States Embassy and Consulates in Japan (2006). Fact Sheet on the U.S. Nuclear Powered Warship Safety. Embassy of the United States. Retrieved 3 May 2017 from http://japan2.usembassy.gov/e/p/2006/tp-20060417-72.html

28 People in HROs that exhibited a *reluctance to simplify* took nothing for granted, always insisting on facts to confirm the true state of a technical process or any of its components. People are encouraged to make fewer assumptions, to notice more, and to ignore less.

29 Weick, K. and Sutcliffe, K. (2007). *Managing the Unexpected: Resilient Performance in an Age of Uncertainty* (2nd edn). San Francisco, CA: Jossey-Bass (pp.10–12, 37–38).

30 Howlett, H. (1995). *The Industrial Operator's Handbook: A Systematic Approach to Industrial Operations*. Pocatello, ID: Techstar (p.45).

31 U.S. Congress. House of Representatives, Committee on Science and Technology (96th Congress, 1st session, 24 May 1979). Statement of Admiral H. G. Rickover, USN, on the Philosophy and Approach of the Naval Reactors Safety Program before the Subcommittee on Energy, Research, and Production (pp.7–8, 18–21).

32 Weick, K. and Sutcliffe, K. (2007). *Managing the Unexpected: Resilient Performance in an Age of Uncertainty* (2nd edn). San Francisco, CA: Jossey-Bass (p.46).

33 U.S. Congress. House of Representatives, Committee on Science and Technology (96th Congress, 1st session, 24 May 1979). Statement of Admiral H. G. Rickover, USN, before the Subcommittee on Energy Research and Production (p.12).

34 Vaughn, D. (1996). *The Challenger Launch Decision: Risky Technology, Culture, and Deviance at NASA*. Chicago, IL: University of Chicago (p.409).

35 Woods, D. (2006). Essential Characteristics of Resilience. In Hollnagel, E., Woods, D. and Leveson, N. (Eds.). *Resilience Engineering: Concepts and Precepts*. Aldershot: Ashgate (pp.29–33).

36 Viner, D. (2015). *Occupational Risk Control: Predicting and Preventing the Unwanted*. Farnham: Gower (p.198).

37 Leveson, N. (2011). *Engineering a Safer World: Systems Thinking Applied to Safety*. Cambridge, MA: MIT Press (p.90).

38 Hollnagel, E. (2008). Safety Management—Looking Back and Looking Forward. Hollnagel, E., Nemeth, C.P. and Dekker, S. (Eds.). *Resilience Engineering Perspectives: Volume 1: Remaining Sensitive to the Possibility of Failure*. Farnham: Ashgate (pp.63–77).

5 CRITICAL STEPS

> What activities, if performed less than adequately, pose the greatest risks to the well-being of the system?
>
> Karl Weick and Kathleen Sutcliffe[1]

> ...effective performance requires both that people can avoid things that go wrong and that they can ensure that things go right.
>
> Erik Hollnagel[2]

Most errors have no consequences, and because of their trivial nature, most errors occur without our knowledge. However, what if an error is unacceptable? What if people's lives are at stake? What if failure really is not an option, as was the case in the following incident?

Near hit: "Man says hold the cheese!"

Several years ago, a West Virginia man in his early 20s said he ordered two hamburgers without cheese at a fast-food drive-through window. Less than an hour later, medical personnel at a local emergency room fought for control over the allergic reaction threatening the man's life. The man was deathly allergic to dairy products and had taken a bite of a hamburger smothered in cheese.[3]

Though the man said he clearly specified "without cheese," it's conceivable during the hectic dinner hour that the order taker heard "with cheese." The customer reportedly said he told the order taker through the intercom and then two workers face-to-face at the pick-up window that he couldn't eat cheese. "By my count, I took at least five independent steps to make sure that thing had no cheese on it," the man said.

After receiving his order, he and his friends departed, assuming he had received the food prepared as requested. They settled down in a dark room to watch a movie. The man took one bite and started having a reaction. His friends rushed him to a hospital where he was treated successfully, though he came close to dying.

Where was the point of no return? As soon as the man took the bite of his sandwich, there was no turning back—the damage was done—his body reacted immediately, experiencing a mild form of anaphylaxis, a rare, potentially life-threatening reaction that impaired his breathing.

The man filed a lawsuit against the restaurant, seeking damages on two counts of negligence, one count of intentional infliction of emotional distress and one count of punitive damages. The lawsuit was dismissed.

A risk-based, yet efficient approach

Thirty-year-old Jill arrives for work well rested after a good night's sleep. She's conscientious, enjoys good health, and has strong family support. "Personal problems" do not weigh her down. In short, Jill is well-trained, mentally alert, physically fit, and faces minimal emotional distractions—an ideal worker. For illustration purposes, let's assume that Jill is 99 percent reliable for the task she is given when she arrives at work.

Jill's boss assigns her a task that consists of exactly 100 actions. (Admittedly, this is a contrived work situation, but there is a point to it.) Assuming the working conditions for every action are the same throughout the job—the chance for success is the same for step 100 as it is for step 1—what is the likelihood that Jill will perform all 100 actions without error?

Jill's performance is a simple probability calculation. The chance for success on step 1 is 0.99 (the action will likely be performed correctly 99 times out of 100 attempts). The chance for success in step 2 is the same, 0.99, and the chance for success on step 3 is—you guessed it—0.99, and so on. The mathematical equation for the probability of successfully completing all 100 actions without error is this:

$$p100 = 0.99 \times 0.99 \times 0.99 \ldots 0.99^{100} \cong 0.3660 \ or \ \approx 37\%$$

It may astound you that the chance of performing just 100 actions without error is only 37 percent for someone who is 99 percent reliable.[4] There's a much better than 50–50 chance that Jill will do something wrong along the way. The news is better if a person's reliability is at the top of the nominal human reliability scale—99.9 percent—but even then, the probability of successfully completing all 100 actions without error improves to just 90 percent. That still equates to a 1 in 10 chance of erring at some point in the 100-step task. A mistake at some steps may not matter, but likely there are other steps at which a mistake could cause a serious problem. The question is which action(s) absolutely have to go right the first time, every time? These are the points in the task at hand, which Jill and her boss should be aware of in order to successfully complete the 100-step task without experiencing serious injury, loss, or damage.

Human fallibility casts a shadow of uncertainty over all high-hazard work. It is impossible from a human perspective to be vigilant 100 percent of the time, to be fully alert, completely informed, and always rational—especially for long periods of time. The human tendency to err is not a problem unless it occurs in sync with substantial transfers of energy, movements of mass, or transmissions of information.[5] When performing work, people are usually concentrating on accomplishing their immediate production goal—not necessarily on safety.[6] If people cannot fully concentrate on safety 100 percent of the time, then when should they be fully focused on safe outcomes? I suggest while performing a CRITICAL STEP.

CRITICAL STEP—a human action that will trigger immediate, irreversible, and intolerable harm to an **asset**, if that action or a preceding action is performed improperly.

Do you need a parachute to sky dive? Literally, no. However, you definitely need a parachute to sky dive twice. Together, the skydiving example, although humorous, and the real-life close call with death for the man consuming a cheeseburger convey the fact that some things have to go right the first time, every time. You'll recall from Chapter 2 that a **touchpoint** is a human interaction with an **asset** or **hazard**, directly or indirectly, such that it changes its state. Some **touchpoints** involve substantial transfers of energy, movements of mass, or transmissions of information that will trigger serious damage, loss, or injury, if not performed under control. These **touchpoints** of no return are what I refer to as CRITICAL STEPS. Though the occurrence of CRITICAL STEPS is a normal and essential part of any production activity, **events** occur at CRITICAL STEPS, if you're not careful.

Notice that the central idea in the definition of a CRITICAL STEP is harm to something of importance—an **asset**. As a reminder, an **asset** is anything of substantial or inherent value to an organization. **Events**, by definition, involve some level of harm to one or more **assets** deemed unacceptable to the organization's members.

In our skydiving example, the all-important **touchpoint** is leaping through the open door of the aircraft—the point of no return. In industry, all work involves a series of **touchpoints** intended to produce an output. Basically, procedures spell out a series of **touchpoints** organized in a particular sequence. But, which **touchpoints** do you pay attention to—which ones absolutely must go right?

> "I have noticed that even those who assert that everything is predestined and that we can change nothing about it still look both ways before they cross the street."
>
> —Stephen Hawking
> Theoretical physicist

You cannot be completely thorough from a safety perspective, and still stay competitive. And, you cannot operate with complete efficiency, if some of your scarce resources are redirected to safety.[7] It is practically impossible to be thorough for every action in a task. However, from a business perspective, efficiency and timeliness are important economic factors for the organization, because there are only so many resources available. Tradeoffs are usually necessary to meet deadlines. But, you always want to meet your commercial deadlines and commitments safely and with the required quality. There is a middle ground you must navigate to accomplish both safety and profitability goals during work.

CRITICAL STEPS offer both an efficient and a risk-based approach to high-hazard work. You have the option to expedite those portions of an operation that has little risk to safety and business accomplishments, but you must slow down for those that do. Identifying and controlling CRITICAL STEPS helps you navigate the safety/production space more optimally. Incorporating the concept and practices of CRITICAL STEPS into your operations has as much to do with productivity as it does with safety. By isolating the more relevant and important human risks, identifying and exercising positive control of CRITICAL STEPS enhances both safety (thoroughness) and productivity (efficiency).

Safety is better considered a process rather than a condition—something the people of an organization do rather than something the organization has.[8] In light of this perspective, managing the occurrence and performance of CRITICAL STEPS is something you do to make sure safety happens to protect **assets** from harm. Managing CRITICAL STEPS requires RISK-BASED THINKING. Managing CRITICAL STEPS enables people to: (1) anticipate what has to go right, (2) pay attention to the right things that alert the worker to impending harm, and (3) respond (adapt, if necessary) in ways to retain positive control of transfers of energy, movements of mass, or transmissions of information, and to minimize the damage or loss should control be lost.

Recognizing a CRITICAL STEP

The concept of CRITICAL STEPS harnesses the abilities to anticipate—to create foresight, to monitor what has to go right, and to respond in ways to preserve positive control of hazardous processes. But, would you know one if you saw one? What criteria would you use to conclude that a procedure step or other human action is a CRITICAL STEP? To help you accurately identify a CRITICAL STEP, it is necessary to know its attributes. Although the context of these actions varies greatly depending on the technology and work environment, CRITICAL STEPS share certain common characteristics, as listed below:

- *Pathway.* A **pathway** exists for a substantial flow of energy, movement of mass, or transmission of information between an **asset** and one or more **hazards**—usually associated with normal work processes. **Hazards** are poised to be released with either one human action or one material failure. For example, the open door on an aircraft cruising at 13,000 feet offers a **pathway** for a skydiver.
- *Touchpoint.* CRITICAL STEPS always occur at **touchpoints**, where human beings interact with an **asset** or **hazard**. For example, depressing the exposure button on an x-ray machine. Overexposure to radiation can cause serious injury to a patient if the machine settings are not set correctly before taking the picture.
- *Certainty.* If control is lost, there is complete assurance that harm triggered by the action will occur—it is sure to happen after the error. For example: touching a hot burner on a stovetop with your unprotected hand will cause pain and injury. Pulling the trigger on a loaded, fully functional firearm will discharge the bullet in the chamber.
- *Immediate.* The occurrence of harm happens faster than an individual can humanly respond to avoid the consequence.

- *Irreversible.* The consequences of the action cannot be reversed. Losing control at a CRITICAL STEP will cause an **asset** to exceed a critical safety parameter—past the point of no return. CRITICAL STEPS do not have an "undo" button. No means exists to restore the **asset** to its previous, unharmed state, or the worker has no means to regain control of transfers of energy, mass, or information once initiated. You cannot unring a bell.
- *Intolerable harm.* Significant injury, damage, or loss is realized. The severity of an outcome is deemed intolerable—depending on what the managers consider important to safety, quality, the environment, and production. The level of severity must be defined for every **asset**.

Not every action that has a point of no return is a CRITICAL STEP. Just because an action cannot be "undone" does not, by itself, constitute a CRITICAL STEP. The designation of an action as a CRITICAL STEP depends on the harm that will be experienced after the action. For example, pressing a selection button on a snack machine, though irreversible, is not a CRITICAL STEP. When you press the button, the snack is immediately transferred to the pickup bin. Yes, it is irreversible (you don't have access to the machine to replace the chocolate bar on the rack). But if executed improperly, the harm is not intolerable. To be useful, the concept of a CRITICAL STEP must be reserved for: (1) what is truly important to the life and health of workers and the public, (2) the proper functioning of the plant and its equipment, (3) the quality of goods and services, and ultimately, (4) the economic survival of the organization.

The following list offers a few examples of actions that satisfy the definition and criteria of a CRITICAL STEP:

- operating a circuit breaker on the power grid that could interrupt electricity to a hospital;
- entering a confined space (where the atmosphere could be deficient in oxygen);
- making an incision during surgery;
- pulling a fuse or an integrated circuit (IC) card from a digital control system;
- taking hold of a bare electrical cable or wire;
- breaching a pressure boundary on a high-pressure system (such as a pipe flange or manway cover on a tank);
- touching the shaft of an operating (rotating) pump;
- operating a process control valve that could interrupt the flow of cooling water to a heat exchanger;
- leaping out of the door of an airborne aircraft while skydiving, or leaping off a platform while bungee jumping;
- clicking "Send," "Submit," "Start," or "Enter";
- speaking potentially embarrassing information in public;
- driving your car into an intersection on a public highway;
- walking across a street (entering any *line of fire*).

In every case, there is either a transfer (or interruption) of energy (electrical, mechanical, heat, etc.), a movement of mass (solids, liquids, or gases), or a transmission of information (data, information, software, signals, authorizations, etc.) that will cause immediate, irreversible, intolerable harm to an **asset**, if the performer loses control or if preceding conditions are not properly established.

Warning! CRITICAL STEPS *are always critical*—whether the **hazard** is present or not. Pulling the trigger on a firearm—loaded or not—is always a CRITICAL STEP. You may have forgotten to eject the round in the chamber. Without a clear understanding that critical is always critical—even if nothing has ever gone wrong before—front-line workers will tend to take safety for granted. They will progressively ease pressure on themselves to stay alert to the potential harm of a CRITICAL STEP gone wrong.

Origins of the CRITICAL STEP concept

The idea of CRITICAL STEPS originated with the handling of nuclear weapons. Understanding which actions really matter was significant because no one wanted to detonate a nuclear bomb by mistake. An uncontrolled high-energy yield would definitely ruin your day—for other workers and likely for people miles around. Nuclear weapons are dismantled at the U.S. Department of Energy's (DOE) Pantex facility near Amarillo, Texas, where "much ado" is paid to CRITICAL STEPS. The facility originally crafted the following definitions:[9]

- *Hazardous step*—"a procedure step that, if performed incorrectly, has a 'potential' to 'immediately' result in a dominant high-energy detonation…"
- CRITICAL STEP—"a procedure step, if skipped or performed incorrectly, that will increase the 'likelihood' of a high-energy detonation … at some later step in the procedure."

You'll note that the distinction between a hazardous step and a CRITICAL STEP lies in the timing and likelihood of when the harm is realized. At Pantex, a hazardous step was considered more serious than a CRITICAL STEP. Later, though, the nuclear industry adopted the phrase "CRITICAL STEP" as the more serious term because it connotes a greater sense of dread and urgency than the word "hazardous." The definition of CRITICAL STEP at Pantex is more in line with what I describe in the next section as a "Risk-Important Action"—still an important concept but not critical in the sense of immediacy. To point out another source: the U.S. Department of Defense (DOD) uses the concept of a "safety critical function." This is a "function, which if performed incorrectly or not performed, may result in death, loss of the system, severe injury, severe occupational illness, or major system damage."[10]

"Famous last words" that demonstrate a wrong attitude toward CRITICAL STEPS:

- "That's odd."
- "Nice doggy."
- "Look Ma, no hands."
- "What does this button do?"
- "Are you sure the power is off?"
- "Don't worry. I've done this a hundred times before."
- "Hey, y'all—watch this!"
- "There's only one way to find out."
- "Trust me—I know what I'm doing."
- "Oops!"

As you can see, both the DOE and DOD definitions suggest a probability as to whether or not an unwanted outcome could result from the step. However, human beings are notoriously inept at estimating probabilities. We usually underestimate the likelihood of an occurrence. That's why I describe CRITICAL STEPS in more concrete terms that minimize doubt as to what can happen. Terms such as "may," "potential," and "likelihood" tempt people to rationalize their sense of control—especially in the throes of production pressures. You and your organization are better off adopting more concrete terms that keep the red lights flashing in the minds of those who perform CRITICAL STEPS. Front-line workers must know: (1) what absolutely must go right (anticipate potential problems), (2) what to pay attention to (actions and critical parameters to monitor their activity), and (3) what to do before they start their work (respond correctly to their job requirements). Does this resemble something addressed earlier? RISK-BASED THINKING?

Too often, people are admonished to "think safety" and to "pay attention," but human beings find it difficult to fully attend to safety and production at the exact same time.[11] Though inspirational, they are marginally effective.[12] When performing tasks, people usually default to accomplishing their immediate production goals, not to safety. Talking about CRITICAL STEPS adds specificity to their understanding of safety, avoiding generalities. The way front-line workers talk about their work either keeps them alert to the dangers at CRITICAL STEPS or allows them to feel dangerously complacent without realizing it.

Personal experience: a walk near the edge

One of my favorite vacation destinations is anywhere on the Continental Divide Trail, which traverses north and south through the Rocky Mountains of the western United States. Once there, backpacking and day hikes are my preferred modes of travel. Once on a hike through Glacier National Park in Montana, I made a profound observation regarding my awareness of **hazards** and my choices to keep me safe from harm in the outdoors. I was performing a "series" of CRITICAL STEPS and didn't realize it!

An especially stunning view literally stopped me in my tracks, and I had to simply stare for a few minutes to drink in the vista before pulling out my camera to record the moment. I continued to walk slowly with my eyes on the horizon of snow-covered peaks. The trail, consisting of gravel strewn treadway, was carved into the side of a mountain. On one side of the trail, a solid wall of granite extended up and out of sight. Roughly three feet away, on the other side of the treadway, there was a precipitous drop-off, at least 200 feet straight down. I had been strolling, and was now standing, along the edge of a sheer cliff. Suddenly, it became very real to me that very little separated me and my enjoyment of this pristine vista from certain death. I had wandered along this cliff for several yards, absorbed by the view, before I realized the danger. Literally, I was one CRITICAL STEP away from certain death.

A simple slip or mis-step would bring a gruesome end to my holiday and grief to my family. Knowing the consequences of a mis-step nudged me away from the edge. After verifying my footing on the treadway, I continued to enjoy the view. Care for my life and the desire to return safely to my family made me more conscious of my footing and my surroundings during the rest of the hike.

Risk-Important Actions (RIAs)

There is another set of actions, preceding CRITICAL STEPS, that establish the pre-conditions required for work both to add value (+) and to protect **assets** from harm (-) during operations. Human actions that create **pathways**, affect the margin of error, or influence an asset's defenses are called Risk-Important Actions (RIA). A RIA includes any one or more of the following human actions that:

- create **pathways** for the transfer of energy, movement of mass, or transmission of information that expose an **asset** to a **hazard** (for the conduct of work);
- influence the number or effectiveness of defenses that protect **assets** (barriers and safeguards);
- impact the ability to maintain positive control of the release of energy, mass, or information (**hazards**) at CRITICAL STEPS;
- verify conditions important for safe operation at CRITICAL STEPS (to avoid land mines).

Preparing a hamburger with cheese is a RIA regardless of who orders it. Wrapping a hamburger with a distinctively-colored paper to denote it as a cheeseburger is a RIA. Placing the hamburger (with cheese) in the hands of a customer possibly allergic to cheese is a RIA. A customer allergic to cheese checking that the sandwich does not contain cheese is a RIA. RIAs are commonplace. For the examples listed below, notice that in every case, an **asset's** condition does not change—there is no substantial release of energy, mass, or information, and the action is reversible as long as the related CRITICAL STEP is not performed.

- Checking that the atmosphere of a large tank is safe for occupancy before entering it to inspect the internal structure.
- Verifying the light switch is off before disconnecting the electrical wires to a lighting fixture in order to install a newer one.
- Adding new oil to an engine during an oil change before starting the engine.
- Donning and securing a parachute harness before exiting the aircraft.
- Stopping your vehicle at a STOP sign before entering an intersection.
- Donning a hard hat, protective eyewear, and gloves before entering a construction zone.
- Securing the lanyard of a fall protection harness after reaching an elevated work location before starting work.
- Closing a water supply valve before disconnecting the fill hose to a toilet tank.
- Venting (depressurizing) a tank before removing the bolts to replace its fluid level sight glass.
- Removing car keys from the ignition and securing them before closing a locked car door.

In any operational process, **pathways** for the creation of value (or harm) are established earlier by preceding actions—where energy, mass, or information are poised for immediate use. Front-line workers usually establish these conditions by one or more preceding procedural actions in the work process. These are RIAs. RIAs establish **pathways**—the conditions for value creation or value extraction.

However, a RIA is a double-edged sword. RIAs create safety and danger. They not only include actions that create the conditions for harm for **assets**, but they also ensure safe outcomes when performing a CRITICAL STEP. Yes, RIAs create **pathways** for harm, but they also involve establishing conditions needed to perform the CRITICAL STEP safely. Table 5.1 lists several RIAs important to skydiving—some associated with danger and others related to safety.

Notice that RIAs precede the related CRITICAL STEP. Every CRITICAL STEP must be set up by one or more earlier actions that create the conditions for value (+) or harm (-); remember, work requires energy. There is at least one RIA that creates the **pathway**—opening the aircraft door for skydiving. There is usually a lapse of time between a RIA and its related CRITICAL STEP. For example, to discharge a firearm, a bullet must necessarily be inserted into the chamber (RIA) before pulling the trigger (CRITICAL STEP). The fact that RIAs precede CRITICAL STEPS suggests that there is time (slack) to check that proper conditions exist before using the firearm, e.g., making sure the safety mechanism is on and pointing the firearm in a safe direction.

Notice also that RIAs are reversible. That means a RIA can be done, undone, redone, undone, redone, and so on, with no immediate negative consequence to the **assets** involved in the activity—as long as the CRITICAL STEP is not performed. Referring to the list of RIAs for skydiving in Table 5.1, you should notice that every step can be reversed. A parachute rigger can rig the main parachute and then decide to unpack the bag and do it over. The skydiver can don and secure

Table 5.1 The sport of skydiving is inherently risky. Certain conditions must be satisfied before leaping out of the aircraft to survive the descent. The table lists the relevant RIAs for the CRITICAL STEP: stepping out of an aircraft door

Actions taken before exiting the aircraft (the CRITICAL STEP)	
Create danger	*Create safety*
• Error by a rigger while packing the parachute (defect)	• Packing main and reserve parachutes properly (barrier)
• Selecting a parachute more complex than the skydiver can handle (error trap)	• Double-checking parachutes rigged properly before closing the backpack (control)
• Taking off in an aircraft and climbing to approximately 13,000 feet (**pathway** required for skydiving)	• Checking pins and the integrity of gear and lines (barrier)
• Securing the harness improperly (defective barrier)	• Acquiring knowledge and skill to conduct jump properly with the selected parachute (eliminate error trap)
• Opening the aircraft door (**pathway** required for skydiving)	
• Stepping toward the door (reduced margin)	• Donning and securing the parachute harness properly (barrier)

Source: United States Parachute Association. Retrieved from http://www.uspa.org.

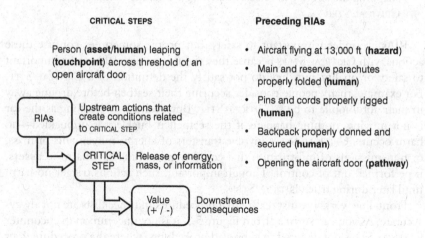

Figure 5.1 The linkage between CRITICAL STEPS and related RIAs is illustrated for the task of skydiving. RIAs always precede their respective CRITICAL STEP and are always reversible—if the CRITICAL STEP has not been performed

the parachute, but then decide not to jump and remove it with no safety-related consequence—as long as the person does not jump out of the door! Figure 5.1 illustrates the linkage between preceding RIAs and their related CRITICAL STEP for the task of skydiving.

To summarize, RIAs possess some or all the following characteristics:

• *Precede its CRITICAL STEP.* Actions, establishing the **pathways** for harm, always occur before the respective CRITICAL STEP.

- *Reversible.* A condition that can be reversed or undone without experiencing damage or loss—little or no transfer of energy, relocation of substances, movement of objects, or communication of important information. There is the option to start over. There is no significant change in the state of **assets**.
- *Slack.* A gap of time exists after a RIA that affords the opportunity to stop work before initiating the transfer of energy, movement of mass, or transmission of information. As long as the CRITICAL STEP is not performed, time is available to think and take recovery actions to preclude unwanted outcomes.
- *Can create latent conditions.* RIAs earlier in a procedure can establish unsafe conditions that may not be detected at the time of performing a CRITICAL STEP—improper rigging of a parachute or incorrect operation of a manual throttle valve that has no position indication. These kinds of actions create land mines in the workplace.
- *Reduced margin of error.* Reduced number of actions to a CRITICAL STEP, such as defeating safety devices or functions prior to performing a CRITICAL STEP. For example: energizing an x-ray machine to a certain power level prior to snapping the shutter to release the energy. There is *no* margin for error at a CRITICAL STEP.

RIAs are indeed important to safety, but people sometimes confuse these actions with CRITICAL STEPS because they rightly perceive RIAs to be important to safety or quality, but they do not satisfy the definition of a CRITICAL STEP. For example, many people consider securing their seatbelt before driving away in their automobile to be a CRITICAL STEP. But, really, is it? As long as the car is not moving, nothing happens if the seatbelt is buckled or unbuckled—no harm occurs. CRITICAL STEPS involve transfers of energy, movements of mass, or transmissions of information that could trigger immediate harm to **assets**, if performed out of control. Though important, such actions trigger no harm until later during true CRITICAL STEPS.

Front-line workers must realize that procedures and checklists are not always accurate. As you can surmise from Figure 5.2, it is not uncommon to encounter CRITICAL STEPS not denoted in approved procedures, where the procedure does not explicitly match the technical and workplace conditions assumed by the author. This is the third possible occasion to identify CRITICAL STEPS. A chronic uneasiness helps front-line workers detect situations in time to make appropriate adjustments or to stop the work and get help before proceeding with the activity.

Warning! Just prior to performing a CRITICAL STEP, one has the opportunity and time to review the outcomes of previously performed RIAs to verify (prove) that conditions are safe for **assets** to proceed with the CRITICAL STEP.

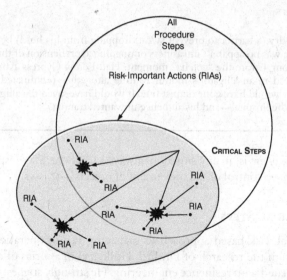

Figure 5.2 RIAs and CRITICAL STEPS in an operational process. Occasionally, CRITICAL STEPS arise that are not included in the procedure because of current work conditions not previously considered by the author of the procedure

Event: when good alligators go bad[13]

Feeling uneasy about wrapping your arms around a 10-foot-long, 350-pound alligator would come naturally to most people—but not Kenny Cypress. The veteran alligator wrestler had astounded thousands of spectators by his antics with the south Florida reptiles when he decided to add a new twist to his circus act: inserting his head inside the open jaws of a wild gator.

In preparing for the trick, Kenny knew how important it was to make sure no foreign object touched the alligator's tongue since even the slightest sensation inside the mouth of an alligator triggers a natural, instinctive reaction for the animal's jaws to snap shut. Kenny had performed this act hundreds of times before, but this time, Kenny made a mistake. After performing the show three times earlier the same day, perhaps the act had become so routine to Kenny that he forgot something. Though he wiped the sweat from the left side of his forehead, he forgot to wipe the right side.

Though there was a helper in the gator pit, the helper did not check or remind him to wipe his brow. On New Year's Day, 1998, a drop of sweat—virtually unavoidable in the muggy Everglades—dripped onto the gator's tongue. Immediately, the gator's jaws crunched down on Kenny's head in a potential death grip. Two hundred horrified tourists watched four men struggle for what seemed like an eternity to pry open the alligator's mouth. Kenny survived but not without scars and some permanent hearing loss.

As a circus act, inserting one's head into the gaping jaws of an alligator—in the *line of fire*, so to speak, is designed to provoke awe, wonder, and terror from a paying audience. The act purposefully created a dangerous situation, which seems to most people to be a foolhardy thing to do, such as working with lions and tigers in a cage. However, the organization, especially Kenny, was aware of the **hazards**, but they were not prepared. No one acted proactively (to double-check) to ensure his forehead was dry (a barrier to prevent sweat dropping from his head) before the key act. There was no apparent uneasiness or ongoing apprehension of things that could go wrong before the "gotcha" moment. Had Kenny Cypress avoided the stick-your-head-in-an-alligator's-mouth routine altogether (eliminated it), his show probably would have gone on just fine. It would have saved the alligator wrestler a trip to the hospital—and his audience unwanted trauma.

Entering the *line of fire* is, in any situation, always a CRITICAL STEP, especially when you do not have control of the movement of objects—or jaws.

Managing CRITICAL STEPS

I believe a practical, risk-based approach to managing **Hu** in operations has been offered through the research of Dr. Erik Hollnagel in a series of articles and chapters associated with resilience engineering. He strongly suggests safety cannot be managed simply by imposing restraints on how work is done—procedures, supervision, automation, and so on. He states clearly that:[14]

> The solution is instead to identify the situations where the variability of everyday performance [behaviors and conditions] may combine to create unwanted effects and to monitor continuously how the system functions in order to intervene and dampen performance variability [error] when it threatens to get out of control.

Managing CRITICAL STEPS involves identifying them for planned work, followed by creating means for the positive control of each one. Anticipating, monitoring, and controlling CRITICAL STEPS and related RIAs are key outcomes of RISK-BASED THINKING. There are three work-related opportunities for identifying and controlling CRITICAL STEPS:

1   Work planning, design, and development.
2   Pre-job briefing just prior to commencing work on the day of the work.
3   During work while in the field.

The last one, during work while in the field, is likely a surprise to the workers and usually involves a flexible response. CRITICAL STEPS are anticipated in the first two opportunities—not so in the third. Front-line workers must have the capacity to recognize and do what is necessary to exercise positive control of CRITICAL STEPS discovered in the field. As emphasized frequently throughout this book, such adaptive capacity depends primarily on technical expertise—the bedrock of RISK-BASED THINKING.

Work planning—CRITICAL STEP MAPPING

Process engineers and procedure writers can be more effective in anticipating and controlling CRITICAL STEPS by using a methodical, repeatable method. One such method, derived from failure modes and effects analysis (FMEA), has been adapted to systematically identify CRITICAL STEPS. This table-top approach, called CRITICAL STEP MAPPING (CSM), methodically identifies **assets**, **hazards**, **touchpoints**, CRITICAL STEPS, their related RIAs, and controls for a specific operation, process, or procedure. A thorough, nuanced review of CSM is beyond the scope of this book, but introducing the idea should help you take a giant step forward in assessing the human risks inherent in your operations. In general terms, a CSM process includes the following high-level phases:

1   *Pinpoint important* **assets** (things of value) to protect from injury, damage, or loss.
2   *Identify significant* **touchpoints** for important work functions (accomplishments).
3   *Identify CRITICAL STEPS and related RIAs.* Compare each **touchpoint** with the definition of a CRITICAL STEP.
4   *Develop defenses* to avoid loss of control and damage to **assets**.
5   *Implement and evaluate controls and barriers.* You'll want input and feedback from front-line workers about the effectiveness of these mechanisms.

CRITICAL STEP MAPPING, used during work planning, helps reveal where safety boundaries exist for the **assets**. These are the "edges of the cliff" figuratively. If, during the analysis, you realize an **asset** will or could undergo a significant change in state, you've likely arrived at a CRITICAL STEP. These are the points in work at which your system needs margin, positive control, and defenses to keep from falling off the cliff.

Pre-job briefing—a meeting of the minds before beginning work

It is helpful to provide front-line workers and their supervisors with structured time to think about their work. Before starting work, workers participate in a pre-job meeting or briefing, sometimes called a tailboard. During this meeting, usually held on the day of the work activity, supervisors and front-line workers review: (1) procedures to familiarize themselves with desired work accomplishments, (2) criteria for success, and (3) what to avoid. Note that the time devoted to a pre-job briefing is part of the job, *not* separate from the job! It is important for the workers to consider explicitly how their actions will affect **assets**. Knowing what to avoid includes identifying, denoting, and controlling CRITICAL STEPS and their related RIAs. Integrating the logic of RISK-BASED THINKING into pre-job briefings and tailboards aids in this awareness. Similar to CRITICAL STEP MAPPING , an adaptation of an approach commonly used in the nuclear power industry to preview a work activity is RU-SAFE:

1   *Recognize* **assets** important to safety, reliability, quality, and production.
2   *Understand* intrinsic operational **hazards** to each **asset** in light of relevant lessons learned from previous incidents.
3   *Summarize* the CRITICAL STEPS and related RIAs from the work plan or procedure.
4   *Anticipate* ways of losing control for each CRITICAL STEP, highlighting dangerous error traps.
5   *Foresee* worst-case consequences for each **asset**, after losing control at each CRITICAL STEP, asking "What if…".
6   *Evaluate* the controls and barriers needed at each CRITICAL STEP, including contingencies, STOP-work criteria, and communication methods.

Individuals can preview their work using the RU-SAFE checklist not only during work preparation, but also during the work-planning process, procedure-development process, and even during the work in the field. It can help anyone think through one's actions before performing them—anytime, anywhere. However, the effectiveness of RU-SAFE depends greatly on the user's technical expertise, without which the worker may not recognize the limits of safety for the **assets**—the edges of the cliff.

Even when no apparent CRITICAL STEPS exist, people should still adopt an attitude of mindfulness and wariness toward their work—a chronic uneasiness—especially during any activity that involves transfers of energy, movements of mass, or transmissions of information during operations. There may be land mines hidden in the woodwork. This is when self-checking is most useful.

During work—performing a CRITICAL STEP

Performing a CRITICAL STEP releases stored energy, initiates the movement of solids, liquids, or gases, or transmits critical data or information. If the RIAs were performed properly, conditions exist that allow the performance of the CRITICAL STEP with no surprises—successful accomplishment of work! Performing CRITICAL STEPS should be anticlimactic all the time.

The four standard rules of gun safety provide a valuable analogy to how workers should approach any work containing a CRITICAL STEP:[15]

1   Always treat a firearm as if it is loaded. Always, until proven otherwise. Keep the firearm unloaded until ready to use.
2   Always point the firearm in a safe direction. Never point the muzzle of a firearm at anything you are unwilling to destroy.
3   Always know your target and what is beyond it (see rule no.2).
4   Always keep your finger off the trigger (outside the trigger guard) until you are ready to shoot. Always.

The preceding series of rules offers insight into the stop-and-review moment before performing a CRITICAL STEP. Although placing your finger inside the trigger guard on the trigger is a RIA (pulling the trigger is the CRITICAL STEP), the time to review is between steps 3 and 4. Usually, there is a brief moment between placing your index finger inside the trigger guard and pulling the trigger. A check here will determine whether it's safe to place your finger on the trigger and pull it. Correctly executing RIAs determines whether someone later gets "killed in action."

When about to perform a CRITICAL STEP, it is important to pause for a moment to consider the situation using a **Hu** tool called self-checking. Self-checking involves stopping the flow of work momentarily to help the performer collect his/her thoughts—to think—to concentrate (pay attention) on what is about to happen. Any action involving the transfer of energy, the movement of mass, or the transmission of information could trigger harm if control is lost.

Self-checking is one of the most effective **Hu** techniques an individual can use to exercise positive control while performing a CRITICAL STEP.[16] Remember, positive control, simply stated is "what is intended to happen is what happens, and that is all that happens." Before acting, the performer thinks explicitly about the intended action, its control, the **asset** and desired accomplishment. The performer ensures safety exists for the **asset** by verifying that proper conditions exist. Self-checking also preserves attention during and after an action. Self-checking's effectiveness depends greatly on the performer's grasp and understanding of the transformation process—his/her technical knowledge.

If uncertain, the performer resolves any questions or concerns before proceeding. When the performer believes a situation places himself/herself, a co-worker, the product, the equipment, or the environment in danger or at risk, he or she STOPS the work, and gets help. STOP means terminating all transfers of energy, movements of mass, and transmissions of information. STOP-work criteria should be explicit, having been discussed previously in the pre-job briefing. Regardless, if there is any doubt, there is no doubt, STOP! *Never proceed in the face of uncertainty!* Once the performer is satisfied that safe conditions exist, he/she performs the CRITICAL STEP—the right action on the right component at the right place and time—while monitoring the **asset's** critical parameters regarding its change in state. The mnemonic, STAR (stop, think, act, review), helps the user to recall the thoughts and actions associated with the self-checking **Hu** tool.[17]

1   *Stop.* Just before transferring energy, moving mass, or transmitting information, *pause* to:

    a.   Focus attention on the **asset(s)** and **hazards(s)**.
    b.   Eliminate distractions.

2   *Think.* Understand what will happen, especially to **assets**, when performing the action.

    a.   Verify the action is appropriate, given the status of the **asset(s)**; understand the **pathway** for the transfer of energy, movement of mass, or transmission of information.

b. Know the critical parameters of the **asset** to monitor, how they should change, and expected result(s) of the action.

c. Consider a contingency to minimize harm if an unexpected result occurs.

d. If there is any doubt, STOP, and get help. Apply STOP-work criteria.

3   *Act.* Perform the correct action under control on the right component.

a.   Without losing eye contact, read and touch the component label.

b.   Compare the component label with the guiding document.

c.   Without losing physical contact, perform the action.

Note: It's common to announce the impending performance of a CRITICAL STEP to others nearby in clear, unambiguous terms, such as:

○ "Clear!" (before shocking a patient in cardiac arrest with a defibrillator).
○ "Fire in the hole!" (before detonating explosives).
○ "Dive, dive" (an alert announced to the crew of a submarine before submerging the ship).

4   *Review.* Verify the anticipated result obtained.

a. Verify the desired change in critical parameters of the **asset(s)**.

b. Perform the contingency, if an unexpected result occurs.

c. STOP work, if the criteria are met, and notify a supervisor or those with the expertise.

Is 99 percent good enough?

What is excellence? Excellence is always described in relative terms as possessing an outstanding quality or superior merit; remarkably good (as compared with others). Most people do most things right most of the time. Recall that nominal human reliability tends to hover between 99 and 99.9 percent. But is 99 to 99.9 percent really good enough? Moreover, is 99.9 percent good enough for CRITICAL STEPS? I think not.

When nothing significant is at stake, 99 percent is fine—great, in fact. Around the house, for instance, most people will never have a serious problem by performing at that rate of reliability. This performance level is fine if the person is simply taking care of household tasks—setting the table, brewing coffee, brushing one's teeth, washing dishes, making the bed, repairing a fence. However, when an error in performing a specific action can cause serious harm to something of value, then 99 percent is certainly not good enough. For example, walking down a straight flight of stairs while carrying a newborn or toddler with two arms. This action *must* go right the first time and every time. *Error-free performance is the only acceptable standard at CRITICAL STEPS.*[18] Indeed, failure really is not an option. A stumble near the top of the stairway could cause you to drop the child, leading to a potentially tragic outcome. Which mistakes do you not dare make at home?

Human error cannot be prevented 100 percent of the time. But when it comes to doing the right thing and doing the right thing right, such as a CRITICAL STEP, errors are to be avoided every time. We expect our employer to honor their obligation to compensate us with the correct amount of funds every pay period, and we expect our banks to account for every penny of it. We expect surgeons to perform flawless medical procedures on our loved ones in need of surgical care. When we travel by commercial aviation, we expect the pilots to fly the aircraft— without error. However, we do expect them to adapt and adjust to the nuances of the situation, such as weather, to protect passengers from harm—do the right thing to the right thing at the right time.

Sometimes, work pressures undermine vigilance and care at CRITICAL STEPS, when speed trumps safety. In his book, *The ETTO Principle*, Dr. Erik Hollnagel points out with regard to the "speed–accuracy trade-off," that *excellence is not good enough*. In critical performance situations, perfection and accuracy are more important than speed.[19] Work *must* slow down to allow front-line workers to think and act circumspectly—to be deliberately certain that **assets** are protected from harm despite production pressures. People prove it is safe to proceed before proceeding—establishing proper pre-conditions via RIAs.

Managers, engineers, supervisors, and workers must all know, understand, and agree on which things absolutely must go right. For non-critical tasks, it may be acceptable to "err" on the side of efficiency. "Get 'r done" can be okay, if the task at hand involves no CRITICAL STEPS. At low-risk **touchpoints** in an operation, I believe it is acceptable to regain lost time or to even cut corners to reduce cost. Who makes that decision? However, bias toward speed and efficiency is *never* appropriate at CRITICAL STEPS and RIAs. The downside risk is simply too great. When 100 percent accuracy is required, 100 percent thoroughness is the only acceptable approach. *Perfection is the ONLY acceptable standard at a CRITICAL STEP!*

Things you can do tomorrow

1   Review the work planning process to understand how high-risk activities are identified, controlled, and denoted in procedures and work packages.
2   For production work scheduled for tomorrow, identify the activities that "cannot fail," and then imagine how they could fail. What can be done to ensure they go without a hitch?
3   Develop an operational definition of a CRITICAL STEP that is relevant to each work groups' work. Verify it satisfies all the common characteristics of a CRITICAL STEP.
4   Using the case study, "When good alligators go bad," ask a work group to identify the CRITICAL STEP(S). Ask the group to judge whether the proposed CRITICAL STEPS satisfy the definition.
5   Just before performing high-risk work activities, ask workers to pinpoint those one, two, or three actions that must absolutely go right the first time, every time. Compare those actions with the definition of a CRITICAL STEP.

Ask workers what they usually do to maintain positive control of those actions. Consider using RU-SAFE to guide the discussion.

6   Build hold points into operating procedures at known CRITICAL STEPS to verify the necessary conditions are established before proceeding (outcomes of RIAs).

7   Do your people know how to self-check? Can you model it for your front-line workers? Encourage people to use the self-checking (STAR) **Hu** tool for CRITICAL STEPS.

Notes

1   Weick, K. and Sutcliffe, K. (2007). *Managing the Unexpected: Resilient Performance in an Age of Uncertainty*. San Francisco, CA: Wiley (p.48).
2   Hollnagel, E. (2014). *Safety-I and Safety-II: The Past and Future of Safety Management*. Farnham: Ashgate (p.146).
3   Anderson, J. (August 10, 2007). Man says, 'Hold the cheese,' claims McDonald's didn't—sues for $10 million. *Charleston Daily Mail*. Retrieved 27 May 2017 from http://www.freerepublic.com/focus/f-news/1879570/posts.
4   Crosby, P. (1984). *Quality Without Tears*. New York: McGraw-Hill (p.76).
5   Center for Chemical Process Safety (CCPS) (1994). *Guidelines for Preventing Human Error in Process Safety*. New York: American Institute of Chemical Engineers (pp.207–211).
6   Hollnagel, E. (2009). The Four Cornerstones of Resilience Engineering, in Nemeth, C., Hollnagel, E. and Dekker, S. (Eds.). *Resilience Engineering Perspectives Volume 2, Preparation and Restoration*. Farnham: Ashgate (p.29).
7   Hollnagel, E. (2009). *The ETTO Principle: Efficiency-Thoroughness, Why Things That Go Right Sometimes Go Wrong*. Farnham: Ashgate (p.28).
8   Ibid. (p.20).
9   Fischer, S. et al. (1998). "Identification of Process Controls for Nuclear Explosive Operations." U.S. Department of Energy (DOE) "Nuclear Explosive Safety Order 452.2A."
10 United States Department of Defense (2000). *Standard Practice for System Safety* (MIL-STD 882D). Washington, DC: Government Printing Office.
11 Hollnagel, E. (2009). *The ETTO Principle—Efficiency-Thoroughness Trade-Off: Why Things That Go Right Sometimes Go Wrong*. Farnham: Ashgate (pp.29–30).
12 Agnew, J. and Daniels, A. (2010). *Safe By Accident?* Atlanta, GA: PMP (p.56).
13 Plaschke, B. (January 28, 1999). Looking for the Ultimate Tail-Gator? *LA Times*. Retrieved 12 April 2017 from http://articles.latimes.com/1999/jan/28/sports/sp-2582.
14 Hollnagel, E. (2014). *Safety-I and Safety-II: The Past and Future of Safety Management*. Farnham: Ashgate (p.121).
15 Adapted from National Rifle Association Gun Safety Rules. (n.d.). NRA Fireman Training. Retrieved 28 March 2017 from http://training.nra.org/nra-gun-safety-rules.aspx.
16 United States Department of Energy (DOE) (2009). DOE Standard: Human Performance Improvement Handbook, Volume 2: Human Performance Tools for Individuals, Work Teams, and Management (DOE-HDBK-1028-2009) (pp.18–19). Retrieved 15 April 2017 from http://www.hss.energy.gov/nuclearsafety/ns/techstds.
17 Ibid. (p.18).
18 Crosby, P.B. (1980). *Quality is Free: The Art of Making Quality Certain*. New York: Mentor Publishing (pp.145–147).
19 Hollnagel, E. (2009). *The ETTO Principle: Efficiency-Thoroughness Trade-Off: Why Things That Go Right Sometimes Go Wrong*. Farnham: Ashgate (p. 52).

6 Systems thinking for H&OP

> If you pit a good performer against a bad system, the system will win almost every time.
>
> Geary Rummler [1]

Events are fundamental surprises. Usually, surprises arise because managers and workers did not fully understand how their organizations worked—how their management systems, technologies, and social interactions influenced individual performance and business outcomes. Because of multiple, layered defenses, high-hazard industrial systems tend to be complex, and no one person can possibly know all there is to know about their true condition. Additionally, because of lean staffing allowances in a competitive marketplace, managers of industrial facilities and their workers tend to be overburdened with multiple responsibilities and seemingly endless tasks. They always seem to have a "full plate." The complexity of human performance in high-hazard operations means that system operations are rarely trouble free as manifested by the following: [2]

- Planners do not and cannot anticipate all contingencies or situations.
- Budgets approximate expenses.
- Equipment wears out.
- Plans are incomplete.
- Wrong replacement parts are used during maintenance.
- Schedules are best guesses, based on experience.
- Procedures include assumptions about initial conditions.
- Technical and software systems (and their components) may be misconfigured or corrupted.
- People miscommunicate.
- People optimistically overestimate capabilities and underestimate resources.
- Plant conditions and people are continually changing.

However, if managers are to reliably protect **assets** from harm during operations and optimize the resilience of their organizations, they must understand how their systems influence people's behavior choices as well as the effectiveness of defenses. This requires that managers understand and practice systems thinking to not only learn successfully but also manage proactively. Systems thinking provides a means of understanding how behavior and results emerge from the systems you design. SYSTEMS LEARNING (described in the next chapter) cannot effectively occur if the management team does not understand and practice systems thinking.

Managers who honestly acknowledge the persistence of human error, accept the fact that their systems—technical, social, and organizational—are not inherently safe. They have to be maintained. Managers in resilient organizations are not seduced by the notion that their facilities, programs, processes, equipment, and tools are without defects—executives, managers, and engineers are also human. Therefore, managers must be intentional and relentless in discovering and correcting weaknesses in their systems. Learning and the continuous search for weaknesses and vulnerabilities must be as valued as the organization's daily production goals.

Human beings tend to simplify and generalize things to reduce their mental workload—a limitation of human nature, equally true of all of us. Managers are no different. Steven Spear, in his book *The High-Velocity Edge*, says "Every time managers make a plan or take an action, it is based on some theory or mental model."[3] To ease their burden, managers depend on tried-and-true mental models to help them efficiently and effectively manage the operations they are trying to control.

In this chapter I address three mental models as frameworks for thinking about human and organizational performance, from a systems perspective—systems thinking. It is important for you, the manager, to understand how behavior, risk, safety, and **events** emerge from the design and inner workings of your social, technological, and organizational systems. Though the technical and engineering design aspects of organizations are important to safety, this book limits its attention to the domain of **Hu** and the organizations people work in.

Note. In this chapter I attempt to "simplify" something that tends to be complex, not just complicated. Some readers may feel it is "over" simplified. To aid managers in better understanding H&OP, I "boil down" the more academic aspects of systems thinking into practical principles relevant to human organizational performance that can be used to manage the risks **Hu** poses to the operation's **assets**.

System basics

Industrial organizations and their operations have become complex systems. Technical systems perform more precise operations at greater speeds with less effort than ever before. Consequently, industrial operations are capable of multiple ripple effects, many of which are hidden from view. However, human capabilities and limitations remain static. Your system comprises several components driving and/or responding to other components in the system, all ultimately influencing workers' choices where work is done. Research has revealed that these system and component interactions can be convoluted, subject to various time frames and multiple factors, and difficult for managers to fully understand what's driving and responding to what.[4]

A system is a set of connected components that interact to perform a specific function or to achieve a specified purpose. Anyone who has spent any time in a technical organization understands that nothing works in isolation—everything is connected. A car, for instance, is a "well-oiled" system. Without giving your actions much thought, you can start it, make it move, and guide it to your destination because the components and processes are joined together within the larger system we call "a car." However, if you disassemble the vehicle—piece by piece—and pile all the parts in the parking lot, you won't stand much chance of driving home that night—because you no longer have a system, even though all the parts are present.

What is the difference between a car and a heap of parts from which the vehicle is assembled? The parts are no longer connected, and hence, they cannot interact. In both cases, the inventory of parts is the same. The crucial difference lies in how the parts relate to each other, performing various functions while receiving feedback from each other in real time to maintain stability and constancy. When the necessary parts are arranged in a certain way, they create a whole system called a car. This car system has characteristics that differ from the constituent parts and is designed to execute a specific function: to transport people. Consistent and efficient outcomes are determined by how well the various components in the system are *aligned* to minimize the effects of variation of individual components, one of which is the driver. Organizations are not much different.

For your organization, the system and the alignment of its components determine how efficiently, profitably, and safely the work of your business gets done. All systems—social, technological, and organizational—share several essential characteristics:

- *All the system's parts must be present for the system to accomplish its purposes.* If a car's wheels are missing, the car will not move. If there is no gasoline in the fuel tank, the engine will not start.
- *The system's parts must be arranged in a specific way to perform its function.* The engine of a car must be able to transfer energy to the wheels through a transmission, or the car, even if it has wheels, still will not move.
- *Systems have specific purposes within larger systems.* People use cars to commute to work in the community's transportation system.
- *Systems use feedback to reach stability.* Feedback provides information regarding the system's performance with respect to its purpose and goals. Speedometers indicate the velocity of the car, which drivers use to guide their decisions on whether to accelerate or slow down.

A rational manager considers these realities every day. Human and organizational performance focuses on the whole work system, not just an individual performing an isolated work activity. To think otherwise would be like supposing that your car will do what you want it to as long as you make sure you have good tires. Good tires may keep you from hydroplaning, but if your steering or braking systems fail, those same good tires may roll you off the road and into a tree. Below, we look at an event where the system failed to protect its most valued assets—passengers.

Later in this chapter, I review the concept of *local factors,* which was introduced in Chapter 3. Managers have substantial leverage over the performance of people in a system through modifications of the conditions people work in (see Figure 3.1). If managers understand how their organizations influence individual performance through *local factors,* how their systems affect the efficacy of its defenses, then there will be better alignment within the system—promoting superior levels of safety, resilience, quality, productivity, and efficiency.

Event: aviation disaster at Dryden

On March 10, 1989, Air Ontario flight 1363 took off from the Dryden, Ontario, Canada, airport in snowy conditions. Although bound for Winnipeg—200 miles away—the flight lasted less than two minutes. At 12:11 p.m. local time, the Fokker F-28 twin-engine jet—carrying 66 passengers and four crew members—crashed in a forest less than a mile from the end of the Dryden runway. Three of the four crew members died in the crash—along with a third of the passengers. A subsequent, three-year investigation cited many factors contributing to the accident, but only the more devastating ones are highlighted here:

- Ice buildup on the wings reduced the amount of lift at takeoff.
- Multiple delays and the desire for punctuality inhibited the occurrence of de-icing.
- While both pilots were highly experienced flying other aircraft, neither had more than 100 hours' experience in the Fokker F-28.
- There was insufficient corporate oversight of the training of pilots—the corporate check pilot had several concurrent responsibilities that distracted him from his training responsibilities.
- The flight attendants were reluctant to question pilots about matters associated with the aircraft. Communication between the cabin crew and the flight deck was not encouraged.
- The airline's maintenance organization was unable to correct a faulty auxiliary power unit (APU) before returning the aircraft to passenger duty. It was allowed to fly without an operable APU—a normal occurrence for the airline.
- Corporate operations dispatched this aircraft with an out-of-service APU to an airport with inadequate ground support equipment. The aircraft could not shut down its engines to support de-icing, which necessitated hazardous "hot refueling" (at least one engine was running while taking on fuel).
- The pilot was reluctant to strand passengers in a town without adequate hotel accommodation—just before a long holiday weekend.
- Another delay occurred just before takeoff, requiring the jet to "hold" on the tarmac. A small aircraft made an emergency landing just as a snow squall passed through Dryden, which allowed more snow to accumulate on the wings. The additional delay further frustrated the pilots, who were keen on adhering to their flight schedules.

In his final report on the Dryden crash, Inquiry Commissioner Virgil Moshansky summarized—in terms of this particular disaster—the very point I am making about managing system weaknesses:[5]

Previous aircraft accident investigations have demonstrated that an accident or serious incident is not normally the result of a single cause, but rather the cumulative result of oversights, shortcuts, and miscues, which, considered in isolation, might have had minimal causal significance.

To assess all of the contributing factors and causes of this accident and to make recommendations in the interest of future accident prevention, the Commission adopted a "systems" approach to facilitate a methodical and thorough investigation. The systems approach reviewed the main components of Air Ontario's transportation system and called for a performance assessment of each of these components, including the national air transportation system. The components of the air transportation system are generally categorized as follows:

- the aircraft crew (including the pilots and the cabin crew);
- the aircraft;
- the immediate operational infrastructure (including airport facilities, navigation aids, weather, and other communications facilities);
- the air carrier;
- the regulator.

… After more than two years of intensive investigation and public hearings, I believe that this accident did not just happen by chance—it was allowed to happen.[6]

System weaknesses allowed conditions and practices to evolve that could be considered "accidents waiting to happen." Several systemic factors identified in the Dryden accident spawned error traps and land mines and degraded the effectiveness of defenses. This tragedy demonstrates the seriousness of the need to understand systems, and why the practice of SYSTEMS LEARNING warrants special attention—before it's too late.[7]

"The consequences of adverse **events** are the natural and inevitable results of the way business is designed and done."[8]

—*William Corcoran*

Systems thinking—blunt end and sharp end

Systems thinking in human and organizational performance facilitates awareness and insight into how front-line worker behaviors and their results *flow* from the context of the system in which the individual performs, that is, *local factors*. Also, systems thinking managers appreciate that their systems influence the organization's defenses (its resilience). As with the previous example of an automobile, systems thinking focuses on the relationships between component parts—between people and the technology, people and organizations, and people and people. Systems thinking examines interdependencies: x influences y and q, y influences z, and z influences x and q. People's choices are influenced by multiple local working conditions, which are, in turn, created by the organization, its management systems, and its values and priorities. People, regardless of where they work in the organization, are part of the system. Though there is free will, there are no free agents. To help you better manage the inherent risk of human error in operations, this chapter addresses some models of performance that will help you think about H&OP.

Warning! "Essentially, all models are wrong, but some are useful."[9] Every model is wrong because it is a simplification of reality. However, some models are useful representations of reality, helping us better understand how things work. They help explain, predict, and understand our environments. Maps are a type of model, but they are always wrong. The map is not the terrain. But good maps are very useful.

Managers who practice systems thinking understand that behavior choices are a product of the influence of multiple system components and processes acting on a person, in the here and now, in the workplace. Systems thinking encourages you to view performance and error as symptoms—behavior choices shaped by the system. Errors that trigger harm have their origins "upstream" in the organization. Individual human errors that trigger **events** are symptomatic of problems deeper in the organization. Managers should acknowledge the culpability of their systems, which are the collective result of many judgments and decisions made over time—drift or wanderings from expectations and the accumulation of fault lines within the organization's defenses (discussed further in Chapter 7). This is why systems thinking is such an important operational worldview for the members of organizations performing high-risk operations. Systems thinking helps managers understand how behavior choices and results emerge from their systems.

All organizations all have two "ends." As illustrated in Figure 6.1, picture an inverted pyramid—with the base or blunt end up and the point or sharp end down. In this image, the organization and its management systems comprise the blunt end, and workers, who do the work of the organization, populate the sharp end. The image of an inverted pyramid represents the funneling effect of multiple, varied, cascading influences (*organizational* and *local factors*) onto the bottom of the inverted pyramid, or sharp end. The weight of all dilemmas, weaknesses, and burdens of the system filters down, pressing hard on the front-line personnel in the workplace, here and now at the sharp end.

The blunt end of the *Systems Thinking Model* comprises people who are removed in space and time from real-time work, comprising executives, policy makers, administrators, lawyers, line managers, engineers, trainers, clerks, human resource specialists, supply chain personnel, and so on. Line managers, in particular, develop plans and expectations, and provide resources, processes, procedures, and incentives that direct and guide the performance of people at the sharp end—*work-as-imagined*.[10] The width of the blunt end suggests that more people work in the organization than at the sharp end.

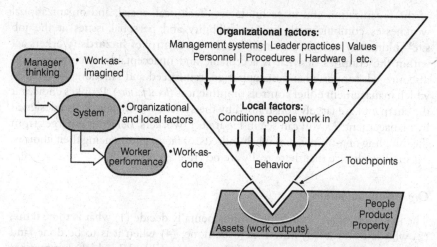

Figure 6.1 Systems Thinking Model: blunt end and sharp end. Behavior and its outputs are the results of multiple organizational factors creating the conditions that people work in. Local factors—products of the organization—explain the difference between work-as-done and work-as-imagined. (Adapted from Figure 1.6 of Reason, J. (1997). *Managing the Risks of Organizational Accidents*. Aldershot: Ashgate (pp. 16–18), and from Figure 1.1 of Woods, D. et al. (2010). *Behind Human Error*. Farnham: Ashgate (pp. 8–9).)

The sharp end of the model represents those individuals who do the organization's "real" work, who add value through the production of products and services—front-line workers and their immediate supervisors. These people come into close contact with **assets** and intrinsic **hazards** in the production processes via multiple and varied **touchpoints**—*work-as-done*. This is where the uncertainty of human fallibility influences work outputs. It is the point of greatest risk—where an **asset** is altered in a way to add value to a product or service or (if something goes wrong) where damage, injury, loss, or death can happen.

The middle layer of the pyramid—*local factors*—offers the greatest insights about performance. *Local factors* occur in the here and now—where and when work happens. Collectively, purposefully or not, they influence the behavior choices of front-line workers during work—at the sharp end. However, *local factors* are part of the system—the outcomes of *organizational factors*—at the blunt end. If there are serious misalignments between *work-as-done* and *work-as-imagined*, the reasons usually reside within the *local factors*—the conditions people work in.

> "You don't have a problem with erratic, unreliable operators.
> You have an organizational problem, a technical one."
> —*David Woods, Sidney Dekker, Richard Cook,
> Leila Johannesen, and Nadine Sarter* [11]

The authors of the book, *Behind Human Error*, suggest strongly that problems that feel like "human error" are not so much individual problems or a cultural one, but a system problem. At the sharp end, workers inherit the flaws and defects of the organization and its technologies. Technical, social, and organizational weaknesses combine with human fallibility and personal issues at the job site—ultimately influencing the risk of losing control of **hazards**. Workers act within the confines of what is acceptable and unacceptable, encouraged and discouraged, supported and unsupported, reinforced and sanctioned—most of which management either controls or influences (*local factors*). People's choices at the sharp end are largely constrained by the system—as imagined and designed by management. As you can see in Figure 6.1, workers' behavior emerges from the mingling of multiple, possibly hundreds, of factors influencing their choices. Again, I emphasize that there really are no "free agents."

Organizational factors

The managers of an organization fundamentally decide (1) what is to be done, (2) who is to do it, (3) how it is to be done, (4) when it is to be done, and (5) where to do it. You are reminded to revisit Table 3.3, which summarizes the various organizational functions managers use to direct and coordinate (influence) the efforts of the organization's members. *Local factors* emerge from the interplay of the various organizational functions—collectively referred to as *organizational factors*. Occasionally, *organizational factors* do not create the desired results. Table 6.1 lists some reasons why.

> "Your management processes are perfectly configured to deliver precisely the product you are currently getting."
>
> —Frederick Taylor
> Author: *Principles of Scientific Management* (1911)

In addition to the adverse *organizational factors* listed in Table 6.1, there are common organizational and system changes that managers and other members of the blunt end should be alert for, as they tend to influence the nature of operational risks encountered by the workforce:[12]

- expansion of production facilities (new equipment and requirements);
- upgrading to new technologies (automation, new knowledge and skill requirements);
- decisions in purchasing (alternative tools, equipment, chemicals, etc.);
- reductions in staffing (a greater reliance on contractors);
- instituting new bonus and reward structures (especially for executives).

Such system changes can have unintended consequences on the behavior choices of the workforce. Weaknesses within the organization are manifested in *local factors*. *Local factors* form the context of **Hu**, whether for good or bad. Basically, managers create and manage *local factors* through *organizational factors*. *Organizational factors*, whether purposefully or inadvertently, create the conditions people work in. This is why managers should adhere to rigorous change-management methods when making "improvements" to the organization.

Table 6.1 Examples of weak organizational factors. The table lists examples of problems at the organization's blunt end that eventually influence performance at the sharp end

- Unexpected effects of future cost-cutting measures (staff reductions, lay-offs) on operations.
- People not developed to assume positions of greater responsibility.
- Unwilling to face and correct small problems.
- Operational impact of changes not evaluated or coordinated.
- Managers not receptive to outside input or using industry operating experiences.
- Lack of a systematic approach to training.
- People don't accept criticism well; pride and defensiveness.
- Adversarial relationship between bargaining unit and management.
- Managers not spending time on the shop floor.
- Some senior personnel too proud to admit things they don't understand.
- Schedule pressure due to ineffective planning, more work needing to be completed with inadequate time.
- Executives relying exclusively on lagging performance indicators looking at past performance; incentivizing **event** rates.
- Uncertain/unclear management objectives and expectations.
- Living with technical problems to save money.
- Poor implementation and follow-through of changes in the organization; lack of accountability.
- Lack of teamwork between managers; infighting; internal competition.
- No consistent corporate communications.
- Operational procedures prepared with mistakes.
- Insufficient tools to accomplish assigned work.
- Complacency—willing to accept the status quo.
- Managers overly optimistic about what they know.
- People do not know how to give feedback.
- Reward system emphasizes short-term gains.
- Because of language differences, some managers are unable to read events reports from foreign facilities, and no budget to translate them.
- Last-minute switching of people assigned to tasks or projects without time to plan or prepare.
- Authoritative leadership style—punitive, degrading, disrespectful.
- Inadequate equipment labeling.

Local factors

The performance of any *part* of a system cannot easily be untangled from the performance of the system as whole. Individual **Hu** cannot easily be isolated from the context in which it occurs.[13] In his book, *Managing the Risks of Organizational Accident*, Dr. Reason describes *local factors* as the "conditions people work in."[14] They set the context for the quality of work, good or bad, and they exist the moment an individual enters the unique workplace to perform a particular task—in the here and now. Everyone has a set of *local factors* that influence his/her choices—including executives and managers. To better prepare you for the next several paragraphs, I suggest you review Chapter 3, especially Figure 3.1, Table 3.2, and Appendix 1 before reading on.

"The frame of reference for understanding people's ideas and people's decisions... should be their own local work context, the context in which they were embedded..."

—Sidney Dekker
Author: *The Field Guide to Understanding 'Human Error'* (2002)

Some *local factors* are more influential than others. In fact, research has shown that the leverage of each cell varies substantially.[16] More than 1,000 representatives of various businesses and industries across the United States were asked, "Improvement in which one of the following six areas (cells) would enable you to do your job better?" and to choose one of the six alternatives that followed the question—the six *local factor* cells. "Leverage" represents the strength of influence the specific category has on performance. Table 6.2 lists the percentage leverage for each cell as revealed by this study.

The factors with the most leverage over performance fall within the power of an organization to make a difference in its results. Managers have more influence over some *local factors* than others. Some factors tend to be more resistant to control than others—especially those associated with the individual, i.e., mental and physical capabilities, personal preferences, motives, and desires. Fortunately, it just so happens that the degree of leverage follows the numbering sequence of the *local factors* cell numbers—managers having the greatest leverage in cell 1 and the least leverage in cell 6. As you can see managers have the most influence on people's behavior choices in cells 1 through 4. The importance of training is not diminished. It is just that people in the workplace did not see training as their greatest need for achieving higher performance levels. External factors, cells 1, 2, and 3, collectively account for almost three-fourths of the performance improvement opportunities. To be most effective in getting people going in the desired direction, emphasize getting cells 1 and 2 right—aligned. Telling people what you want, letting them know how well they are achieving what you want, while giving them the resources to do what you want, will go a long way in accomplishing work safely and reliably.

Table 6.2 Local factors are not created equal. Some have more leverage (influence) on behavior choices than others. In general, the leverage on behavior is ranked by cell (1 through 6)

Task and environmental factors (external to the person)		Human and individual factors (internal to the person)	
1. Expectations and feedback	35.3%	4. Knowledge and skills	10.5%
2. Tools, resources, and work environment	26.0%	5. Capacity and readiness	7.7%
3. Incentives and disincentives	11.3%	6. Personal motives and expectations	6.3%

Source: Dean, P. and Ripley, D. (Eds.). *Performance Improvement Pathfinders: Models for Organizational Learning*. Silver Spring, MD: ISPI (pp.52–55).

Referring to Table 6.3, if the organization is not achieving its desired business results (R_2)—currently achieving undesired results (R_1)—there is misalignment. To understand where changes are needed, the *Alignment Model* (see Figure 3.2) can be used to guide your analysis, using it in reverse order. To improve business results from R_1 (what you're getting) to a new R_2 (what you want), you'll have to alter the product, quality, and/or productivity of work outputs from W_1 to a new W_2. A new set of work outputs (deliverables), W_2, will require a change in behavior from B_1 to B_2. But, if you want to change behavior and sustain it, you'll have to change the context of **Hu**, that is, *local factors*, from L_1 to a new set of influencers, L_2. *Local factors* ultimately influence what people do moment by moment in the workplace. But, to sustain this new performance level, you'll have to change the organization and its management systems from its original set of *organizational factors*, O_1, to a new set, O_2.[17]

Rarely is a new behavior sustained with changes to just one or two cells. Usually, sustained behavior change requires a combination of several influencers from several cells acting in concert.[15] Hence, the need for system alignment.

> *Caution! Local factors* vary from job to job, location to location, day to day, and person to person.

At this point, I want you to make an important observation about the *Alignment Model*. Every element in the model is a static condition except one—behavior. Behavior is an action, where work occurs. The other four elements are static outcomes. Without aligning the organization to create a new set of *local factors* (engineering it in a manner of speaking) to get a new set of behaviors (B_2), sustained, long-term improvements in performance ($B + R$) are unachievable. If you want safety over the long haul, you'll have to tune (engineer) your organization to adapt to changing risk realities in the workplace. Otherwise, "you'll always get what you've always got." Hence, the derogatory phrase "program of the day."

Table 6.3 Change requires realignment. To achieve and sustain a new level of business, R_2, it is necessary to realign the organization to direct people's behaviors in a new way. New behavior choices require a new set of influencers—local factors, L_2, to both enable the new practices, B_2, and inhibit the previous practices, B_1

Work-as-imagined (new behavior choices)	Alignment	Work-as-done (previous practices)
O_2	Organizational factors	O_1
L_2	Local factors	L_1
B_2	Behaviors	B_1
W_2	Work outputs	W_1
R_2	Business results	R_1

Later, in Chapter 8, I explore in greater detail how managers and supervisors can be more effective in managing **Hu**, using the *Alignment Model*.

Complicated vs. complex systems

The primary focus of this book is on managing the risk of human error in complex operations. But, the principles espoused herein can also be used to address complicated work situations. Dr. Ivan Pupulidy, a former member of the U.S. Forest Service, describes the difference between complex and complicated operations this way:

> "I can take apart my motorcycle and put it back together, but I can't take apart my horse and put it back together. Motorcycles are complicated. Horses are complex."[18]
>
> — *Ivan Pupulidy*

Managers are often challenged with human and organizational performance because they think they are operating in complicated environments when they are really operating in complex ones. Machines don't think (at least not yet), but people do. Managers oversimplify the management of people, often thinking of them as simple machines when they are in fact complex. All organizations comprise a group of independent agents—people, each with differing personal goals, motives, and preferences, and they make choices. When working toward a common goal, people want to be efficient and thorough, but they sometimes come to different conclusions about how to accomplish their goals. People tend to adapt to their surroundings, which tends to add complexity to operations. When used together, the *Systems Thinking Model* and the *Alignment Model*, help managers guide people toward safer, more conservative behavior choices in complex, high-hazard environments. Although oversimplifications, the mental models addressed in this chapter can help managers and leaders better understand how to optimize H&OP to take advantage of the system's leverage to influence behavior choices in the workplace. To make that happen, let's take a closer look at the "battery."

The work execution process

The *Work Execution Process* focuses on the sharp end—that part of the *Systems Thinking Model* described earlier in this chapter at the behavior–**asset** interface in the workplace where work is done—at the notch (refer to Figure 6.1). To enhance your ability to manage **Hu** for high-risk work, you need a clear picture of work as a process. It guides the day-to-day management of and interactions between people, **assets**, and **hazards** in the here and now. A process is a repeatable sequence of functions or actions ordered to: (1) accomplish a goal, (2) produce a product, or (3) achieve an end. In this light, you can also see that managing is itself a process: plan, do, check, and adjust—which enhances your RISK-BASED THINKING as the work process manager.

As illustrated in Figure 6.2, the *Work Execution Process* involves three phases: (1) preparation, (2) execution, and (3) learning. When H&OP is better understood, the process integrates RISK-BASED THINKING into each phase of work to help supervisors and front-line workers more consistently control their potential impact on **assets** and environments.

Preparation

Preparation for high-risk work requires a dedicated chunk of time to allow workers and supervisors to get ready mentally. Before engaging in hands-on work, workers carefully review procedures and other related guidance to develop a clear mental picture of current system conditions, desired work outputs, and the plan of how to achieve them. For teams, a dialogue normally takes place. As described in Chapter 5, this dialogue (pre-job briefing) clearly defines what is to be *accomplished* as well as what is to be *avoided*. Workers talk about what must go right and how to avoid harm—who does what, to what, with what, and when. When preparation is complete, there is no ambiguity about:

- what the **assets** are and the potential **hazards** to each (critical parameters);
- the work to be done (**pathways**) and expected work outputs or end states (success criteria);
- CRITICAL STEPS and related RIAs (**touchpoints**);
- proper pre-conditions to ensure control of **hazards** and behavior choices;

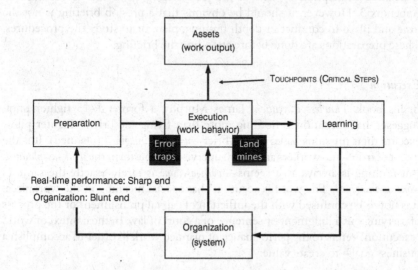

Figure 6.2 The Work Execution Process. Work (transformation) occurs at the sharp end of the organization where front-line workers "touch" things (work behavior). Chances for success are optimized when front-line workers are given time to think (preparation) about what is to be accomplished as well as what is to be avoided for the work at hand. Time for learning after the work is allotted to improve the safety and reliability of future performances

- who does what, in what order, and with what resources, including the methods to accomplish the work under control (positive control);
- communication channels to ask for help when things don't go as planned.

Preparation optimizes the likelihood that people will succeed.[19] The information reviewed in the preparation stage gives each person the ability to visualize their personal contribution to the work at hand. The business purposes of their work must be clear to the workers because knowing specifically what is to be accomplished improves people's ability to more readily recognize when/where they might need to adjust to changing risks in the workplace. Clearly specifying what actions and conditions are expected to ensure desired outcomes, makes people more adept at detecting when something unusual happens.

Preparation helps the assigned field worker know in advance what to do to maintain positive control of specific actions important for safety, especially CRITICAL STEPS. It also gives workers time to think about answers to questions, such as "What if…?" and "What's the worst that could happen?" Unless workers precisely pinpoint the **assets** they will work on, the ability to manage the risk of human error is significantly hampered. To effectively anticipate which errors to avoid during work, it is important for people to know what they are going to touch—alter, change, manipulate, move, etc.—and which **touchpoints** are critical to safety, quality, reliability, and production.

In most cases, pre-job briefings occur on the day of the job, and they usually last between 15 to 30 minutes—depending on the work's risk and complexity. A detailed description of what a pre-job briefing looks like is provided in Appendix 3. However, it should be obvious that a pre-job briefing is not the time and place to conduct in-depth training, plan, or to study the procedures. These preparations are done before attending the briefing.

Execution

In his book, *Flawless Execution*, James Murphy, a former F-15 fighter pilot, suggests that execution is nothing more than flying the brief.[20] Fighter pilots execute their missions based on the brief—*work-as-imagined* (planned). It is the intent of front-line workers and their supervisors to perform the *work-as-planned*. But nothing is always as it seems—*work-as-done* is rarely exactly the same as *work-as-imagined*. The word "execution" has special meaning in the workplace. It is not to be confused with the infliction of capital punishment or the process of carrying out a judgment or sentence of a court of law. In the context of work, "execution" refers to the performance of planned work in order to accomplish a business goal—to create value.

> *Warning!* War plans rarely survive first contact with the enemy. Similarly, *work-as-done* is rarely exactly the same as *work-as-imagined*.

After preparation, workers begin their tasks. They start touching and altering things to accomplish their goals through specific transfers of energy, movements of mass, and/or transmissions of information. At this point in the *Work Execution Process*, it is important for the worker to know in no uncertain terms what the CRITICAL STEP(S) are, why they are critical from a safety standpoint, and how to exercise positive control through the use of specific **Hu** tools. It is at this point when real work is performed that people come into close, physical contact with the organization's **assets**, products, services, and operational **hazards**—where human errors can be the most detrimental to safety, quality, reliability, and production. The performer wants to avoid harm and inefficiencies. A CRITICAL STEP offers the performer the opportunity to pause before proceeding with the action—to verify that safe pre-conditions are established before proceeding with the work—*prove it safe* at RIAs!

Preferably, all CRITICAL STEPS are identified during the pre-job briefing. But, too often, this is not the case. Ideally, work planners and procedure writers identify and denote all known CRITICAL STEPS in work documents. But, being human, they can overlook one or more, given their assumptions about the work situation and their technical understanding of the production process. Similarly, supervisors and front-line workers could conceivably fail to notice one or more CRITICAL STEPS during their individual preparation and the pre-job briefing. Recall that CRITICAL STEPS may surface during chunks of work the procedure writer or planner considers "skills-of-the-craft." Also, work conditions may change or are different from what was originally planned for. Occasionally, because of land mines in the workplace, CRITICAL STEPS show up unannounced. If the front-line worker has a chronic uneasiness—mindful of transfers of energy, movements of mass, and transmissions of information—these situations may be recognized and controlled before acting. Therefore, front-line workers must understand intimately the characteristics of CRITICAL STEPS.

The likelihood of losing control at a CRITICAL STEP can be minimized in several ways. First, structure the work plan to eliminate or minimize the effects of error traps. For example, the timeframe allotted for a job can be scheduled in a way to avoid hurrying. Second, train and develop front-line personnel on the prerequisite technical expertise **Hu** fundamentals. Third, promote a chronic uneasiness. Chronic uneasiness promotes situation awareness and a sensitivity of **pathways** for transfers of energy, movements of mass, or transmissions of information during the work at hand. Fourth, practice the use of **Hu** tools at CRITICAL STEPS and their related RIAs. **Hu** tools improve positive control and promote resilient behavior, enhancing people's capacity to adapt—to make needed field adjustments in response to risks to **assets** that may occur unexpectedly. Finally, managers can improve the chances of safe performance by designing work with slack and flexibility built in. To enhance adaptability during high-risk activities, consider building in the following forms of flexibility:

- time;
- information (availability and accessibility);

- tools (accessibility and usability);
- parts (replacement components needed for vital pieces of equipment);
- expertise (availability technical knowledge, skills, and experience with task at hand);
- other people (availability of extra brains, eyes, and hands).

Recall that the choices people make, in the here and now, are the collective influence of a unique set of *local factors* present for the task at hand. Managers have control of those that have the most leverage over behavior choices—primarily cells 1 through 4 (see Appendix 1).

There is always something that does not go as expected. When the work is done, front-line workers and their supervisors should be given time to think about what they have done.

Learning

Learning depends on feedback, and there are two primary avenues of feedback during work—feedback from workers to managers and feedback from managers to workers. Both are necessary. Work-related feedback (1) provides managers with opportunities to improve management systems among other systemic factors, and (2) presents occasions to improve individual worker performance involved in a work activity. More specifically:

- *Field observations*—the opportunity for managers and supervisors to acquire firsthand information about the effectiveness of work planning and to better understand worker challenges, concerns, readiness, and actual performance.
- *Reporting*—feedback from workers that provides managers with a rich, valid, and fresh source of information about task-specific conditions, procedures, resources, coordination, incentives and disincentives, and their related management systems.

These are two primary sources of learning that managers must take advantage of. More details on each method are provided in the next chapter on SYSTEMS LEARNING.

"Learning" from a RISK-BASED THINKING perspective involves knowing (1) what has happened, (2) what is happening, and (3) what to change going forward. Even though this last phase of work is devoted to learning, you should not presume that learning occurs only after work execution (knowing what to change). Learning occurs in all three phases of the *Work Execution Process*. Learning from previous **events** and the mistakes of others is an expectation during pre-job briefings in the "preparation" phase (knowing what has happened). Feedback about personal performance occurs during field observations by managers and supervisors, and managers, in turn, receive feedback about the functioning of their management systems during the "execution" phase (knowing what is happening). However, when work is completed, in a commercial production environment, it is always tempting to move on to the next task immediately. If so, managers are missing a great opportunity to know more about the effectiveness of their systems.

Post-job reviews (also known as after-action reviews) provide an opportunity for a work team to briefly discuss any significant *differences* (including surprises) between *work-as-imagined* and *work-as-done*—the key differences between preparation (what they planned to do) and execution (what they actually did). Post-job reviews provide workers and supervisors with a few minutes to reflect on and report significant issues (positive and negative) that need management's attention before performing the job or similar jobs in the future.

Report: wolves redirect the Yellowstone River

In 1995, after the reintroduction of wolves into Yellowstone National Park in Wyoming (after a 70-year absence), scientists found that aspen, willow, and cottonwood groves expanded in size and that shrubbery and grasses became more prevalent, especially around rivers and streams. This occurred, not because wolves were killing the elk, but because the elk stopped grazing like domestic livestock. Their feeding behaviors were altered drastically by the presence of wolves. The regeneration of the trees and grasses along rivers stabilized their banks, causing the rivers to become more fixed in their courses—a pleasant, but unexpected result.[21]

Occasionally, the implementation of change in your organization incurs surprise outcomes, both positive and negative. This may be due to a lack of understanding of how the change might impact relevant stakeholders. For example, a frustrated warehouse manager tells you he has no idea how he is going to store the refrigerated stock you just ordered, since the warehouse walk-in cooler only has capacity to store 20 percent of what you ordered. Or, the quality control test lab manager complains when he receives a delivery of 200 raw material samples, increasing their workload 200 percent, and they have no budget for overtime.

An organization is not unlike an ecosystem, such as Yellowstone National Park. The processes and systems that work in harmony one day can lead to unintended results the next—especially if you do not fully understand how your organization works or doesn't work.

Systems thinking—the flow of influence

As you become comfortable using the *Systems Thinking Model*, the *Alignment Model*, and the *Work Execution Process*, you'll begin to practice systems thinking. All three models illustrate the flow of influence through an organization (via various avenues or organizational functions) on the behavior choices of front-line personnel and on their work outputs. As described before, most *local factors* are systemic products at the blunt end of the organization—the system—that collectively influence an individual's behavior during job performance.[22] These conditions press hard upon the sharp end where and when work takes place, which is modeled by the *Work Execution Process*. When work involves high-risk operations, you do not want to leave performance to chance. Manage every stage of the *Work Execution Process* to make sure the work and its CRITICAL STEPS are accomplished safely and reliably, whether by approved procedures or by sound field adjustments.

All three models help managers in two ways. First, they suggest how change and improvement occur. Your change management plans should be consistent with these models as they describe how organizational alignment of the "flow of influence" between the blunt and sharp ends of your operation is necessary in obtaining desired business results through people. Second, tracking the flow of influences that existed before an **event** clearly shows the "causes" of unwanted outcomes. When there is misalignment, *work-as-done* and its outcomes at the sharp end do not match *work-as-imagined* by those at the blunt end. This is normal for the most part, but sometimes serious misalignments emerge unexpectedly, eventually resulting in **events**. Therefore, analyses of **events** should review the flow of influences using the models in reverse order, from business results to *organizational factors*, to reveal systemic misalignments. Corrective actions to stabilize or improve performance and business outcomes, as well as safety and resilience, by necessity, involve fundamental changes to organizational functions to sustain change—the institutional ways you do business. In light of the changes in the course of the Yellowstone River from the re-introduction of wolves into the ecosystem, managers would do well to study how their organizational systems work or don't work—SYSTEMS LEARNING.

Things you can do tomorrow

1   During the next **event** analysis review, ask the analyst(s) to describe the cause model used in the investigation. Check their understanding of how behavior choices flow from the system—*Systems Thinking* and *Alignment Models*. Review the content of cause analysis training that an **event** analyst receives to determine if it includes systems thinking.

2   Conduct a case study of a previous **event** to describe how system misalignments contributed to the consequences and people's behavior choices.

3   Reflect on your mindset and assumptions about workers and management systems when an unwanted outcome or an **event** occurs, or whenever actual work (*work-as-done*) does not match your expectations (*work-as-imagined*).

4   During management meetings, apply the *Systems Thinking Model* explicitly to the issues at hand—how the organizational decisions could influence behavior choices on the shop floor, and how they could influence the integrity of defenses (safety and productivity of operations).

5   Post an enlargement of the *Generic Local factors* (Appendix 1) and the *Systems Thinking Model* in management meeting rooms. Refer to these models as an aid in making important operational decisions.

6   At the next management meeting, ask participants how their operational decisions will impact safety. Ask them how they will avoid harm. Discuss the typical questions you ask people when you are in the operating spaces. Discuss the key messages they send to the workforce.

7   Using a systematic approach to training, develop and conduct manager training on the *Systems Thinking* and *Alignment Models*. Refer to the section on *Training* in Chapter 8. Include case studies to illustrate the power of the mental models on managers' decision-making processes.

8   Post the *Work Execution Model* in workshops or anyplace work planning and preparation occur. Encourage workgroups to improve their preparation for the CRITICAL STEPS of the work and to learn more deliberately from their experiences on the job.

Notes

1   Rummler, G. and Brache, A. (1995). *Improving Performance: Managing the White Space on the Organizational Chart*. San Francisco, CA: Jossey-Bass (p.13).

2   Woods, D., Dekker, S., Cook, R., Johannesen, L. and Sarter, N. (2010). *Behind Human Error*. Farnham: Ashgate (pp.8–11).

3   Spear, S. (2009). *The High-Velocity Edge*. New York: McGraw-Hill (pp.xiii–xx).

4   Hollnagel, E. (2014). *Safety-I and Safety-II: The Past and Future of Safety Management*. Farnham: Ashgate (pp.107–113, 118).

5   Moshansky, Virgil P. (1992). *Commission of Inquiry into the Air Ontario Crash at Dryden Final Report*, Vol. 1, Canada: Minister of Supply and Services.

6   Ibid. (1992). pp.4–6.

7   Viner, D. (2015). *Occupational Risk Control: Predicting and Preventing the Unwanted*. Farnham: Gower (p.48).

8   Corcoran, W. (April 2001). The management system as a root cause. *The Firebird Forum Newsletter*, 4(3) (p.2).

9   Box, George E.P. and Draper, N.R. (1987). *Empirical Model-Building and Response Surfaces*. New York: Wiley (p. 424).

10 Woods, D., Dekker, S., Cook, R., Johannesen, L. and Sarter, N. (2010). *Behind Human Error* (2nd edn). Farnham: Ashgate (pp.8–11).

11 Woods, D., Dekker, S., Cook, R., Johannesen, L. and Sarter, N. (2010). *Behind Human Error* (2nd edn). Farnham: Ashgate (p.239).

12 Viner, D. (2015). *Operational Risk Control: Predicting and Preventing the Unwanted*. Farnham: Gower (p.117).

13 Shorrock, S., Leonhardt, J., Licu, T. and Peters, C. (August 2014). Systems Thinking for Safety: A Whitepaper." Brussels: EUROCONTROL.

14 Reason, J. (1997). *Managing the Risks of Organizational Accidents*. Aldershot: Ashgate (pp.16–18, 121, 223).

15 Patterson, K., Grenny, J., Maxfield, D., McMillan, R. and Switzler, A. (2008). *Influencer: The Power to Change Anything*. New York: McGraw-Hill (pp.76, 259–264).

16 Dean, P. (1997). "Tom Gilbert: Engineering Performance With or Without Training." In Dean, P. and Ripley, D. (Eds.) *Performance Improvement Pathfinders: Models for Organizational Learning*. Silver Spring, MD: International Society for Performance Improvement (ISPI) (pp.52–55).

17 Connors, R. and Smith, T. (2011). *Change the Culture Change the Game*. New York: Penguin (pp.29–66).

18 Personal conversation.

19 Spear, S. (2009). *The High-Velocity Edge*. New York: McGraw-Hill (p.23).

20 Murphy, J. (2005). *Flawless Execution: Use the Techniques and Systems of America's Fighter Pilots to Perform at Your Peak and Win the Battles of the Business World*. New York: Regan Books (p.19).

21 Many thanks to Ms. Luisa Muscara for revealing this report to me. Sustainable Human (February 13, 2014). *How Wolves Change Rivers* [video file]. Retrieved 27 May 2017 from https://www.youtube.com/watch?v=ysa5OBhXz-Q.

22 Van Tiem, D., et al. (2004). *Fundamentals of Performance Technology: A Guide to Improving People, Process, and Performance* (2nd edn). Silver Springs, MD: International Society for Performance Improvement (ISPI) (pp.8–10).

7 SYSTEMS LEARNING

Things are the way they are because:

- managers want them that way, or
- managers tolerate them that way, or
- managers are unaware of them.

William Corcoran [1]

If we cannot have perfect human beings then why should we expect, philosophically, that machines designed by human beings will be more perfect than their creators?

Hyman Rickover [2]

Since 1975, traffic fatalities have actually gone down in the United States, despite the fact that the number of licensed drivers and total miles driven per person increased significantly. The number of licensed drivers in the United States for every motor vehicle crash death increased from approximately 2,900 to more than 6,500 from 1975 to 2013 (see Table 7.1). What changed? Safer cars, better highways, and so on. How about people? Not so much. Substantive improvements in the design of automobiles and the transportation system (highways and infrastructure) explain in large measure much of the reduction in motor vehicle deaths. People, too, have likely become more cautious drivers, and there is greater use of seat belts, especially for children.[3] What's the bottom line? If you want safety, whether individually or system safety, identify and correct weaknesses in the system. Don't rely on "preventing human error."

What is systems learning?

Learning is a key point of leverage in successfully managing H&OP—protecting assets from harm. Learning is demonstrated only when behavior changes whether individually or organizationally. But changes in behavior cannot occur unless there is feedback. Learning starts with discovery and feedback—nothing can change without knowing where you are. Feedback provides the learner

Table 7.1 A comparison of traffic fatalities between 1975 and 2013 with the number of licensed drivers reveals that something other than people's nature changed in the transportation system

	1975	2013
No. of traffic fatalities	44,525	32,719
No. of licensed drivers	129,814,873	212,159,728

Source: Federal Highway Administration, U.S. Department of Transportation. Retrieved from https://www.fhwa.dot.gov/policyinformation/statistics.cfm.

with information that creates an opportunity to change behavior. Consequently, discovery and feedback are key features of sustained high levels of system performance and safety.

Manifestations of system weaknesses—deviations, work-arounds, error traps, mis-configured process systems, at-risk practices, worn-out tools and equipment, trade-offs, and land mines—are all around us. In fact, *the causes of the next **event** exist today*. That should make you tremble with fear. What do you do about it? Find the problems, especially the ones that are happening frequently during high-risk operations, understand how widespread they are and why they exist, and then fix them. Simple, right? In his book, *Controlling the Controllable*, Dr. Jop Groeneweg summarizes the three steps to managing your deficiencies:[4]

1 Detect deviations.
2 Know the extent of these deviations.
3 Manage them.

SYSTEMS LEARNING is that simple. In essence, learning is really a problem-solving process focused on the organization and its management systems—it's a form of organizational adjustment to better match subtle variations in operational risks. SYSTEMS LEARNING does not happen by itself. But, you must purposefully and diligently dig to find the truth and then do something with it.

Learning does not only involve the acquisition of new knowledge. It also demands a response—a correction of behavior. Lessons are not learned until behavior changes and not sustained until organizational and management systems are revised. I define a "lesson learned" as a change in behavior, whether personal or organizational. But, if nothing changes, there are no lessons learned. You have to be diligent and persistent in finding your problems—you cannot let up. Then, and only then, will you be able to correct them, minimizing your time at risk of an **event**. *Detect and correct*.

Learning must not only happen, but it must happen fast. Intense production stresses and marketplace demands can pressure managers to postpone learning as it does take time and resources away from "more important" mission-related operations. Should you wait to learn? Learning late can be deadly and costly. The longer you wait to make needed changes, the longer the "time at risk" and the gamble. Time at risk is the period of time that an unsafe or at-risk condition (land mine or system weakness) exists from the time of its creation to the time the condition is corrected. During this period, an active error or other activities (normal work, violation, or an "act of God") could combine with the unsafe condition to trigger an **event**.

Steven Spear, in his book, *The High-Velocity Edge*, suggests that high levels of performance are achieved through a "relentless pursuit of truth" about the actual functioning of an organization. While highlighting the work of Alcoa Aluminum, he states:[5]

…no team can design a perfect system in advance, planning for every contingency and nuance. Despite all the effort put into up-front design, something will always be overlooked. Alcoa realized they can discover great systems and keep discovering how to make them better. Alcoa gave up on designing perfect processes and committed itself to discovering them instead.

For managers, the first and primary challenge to managing latent system weaknesses is finding them—discovery. Detecting deviations is the most important feature of Systems Learning. Recall that the word latent essentially means hidden. Depending on the technology, tempo, scope of operations, and the range and type of defenses used, among other factors, the number and frequency of interactions can lead to a level of complexity that makes it difficult for a manager to always know what is actually happening with people at any one time. This could make discovery problematic. Because of complexity, you will need multiple and varied ways of detecting weaknesses in your operation. *Detect and correct*.

"If you always do what you've always done, you always get what you've always gotten."

—Jessie Potter

Research study: why systems learning is so important

Human factors specialists at the U.S. Department of Energy's Idaho National Laboratory (INL), commissioned by the U.S. Nuclear Regulatory Commission, explored the changes in risks (i.e., potential severity) associated with active and latent errors to the safety of the reactor cores of operating commercial nuclear power plants in the United States. The government report, NUREG/CR 6753, documented this analysis.[6] INL's human factors specialists analyzed 37 selected **events** that occurred in the commercial nuclear power industry over a six-year period and made the following assertions in the report's Executive Summary:[7]

The results showed that **Hu** contributed significantly to analyzed **events**. Two hundred and seventy (270) human errors were identified in the **events** reviewed, and multiple human errors were involved in every **event**. Latent errors [i.e., errors committed prior to the **event** whose effects are not discovered until an **event** occurs] were present four times more often than were active errors (i.e., those occurring during **event** response). The latent errors included failures to correct known problems and errors committed during design, maintenance, and operations activities.

Most of the latent errors in these **events** manifested themselves as weaknesses with various management systems or flawed or missing defenses. Notice that latent errors (actually latent conditions) were present in every **event** analyzed and were more prevalent than active errors. The manifestations of latent errors were noted in all facets of organizational functioning:

- operations
- design
- design change work practices
- maintenance practices
- maintenance work controls
- procedures and procedure development
- corrective action program
- management and supervision.

The study also corroborates what we already know about latent errors (conditions) in that they accumulate over time, possibly present for many years, until such conditions may combine with local workplace circumstances, equipment failures, or active errors by front-line workers to trigger operating **events**. The report also concluded that *latent errors were the primary contributors to the events studied*; active [errors] by operations personnel were not. The report goes on to conclude, "The errors that contributed most often to plant **events** and caused the greatest increases in plant risk were latent errors." This suggests that the severity of future **events** is more a function of system weaknesses and related weaknesses in defenses than active errors by front-line workers.

Too often, safety is viewed as a problem of the mis-steps of front-line workers—a red herring of problems deeper in the system. The INL study helps substantiate how important it is to find and correct long-standing, often hidden, system weaknesses and how wrong it is to simply "blame the worker" for causing **events**. In most cases when serious harm has occurred, either several defenses had to fail or the margins for safe operations eroded to the point of losing control, which are mostly attributable to a variety of system weaknesses.

Educator and counselor (1981)

The study described here strongly suggests that executives and managers must view SYSTEMS LEARNING as an essential business function, equivalent in importance to managing the financial and production activities of the business. Just as Alcoa realized, you cannot always prevent latent weaknesses and adverse workplace conditions from emerging, but you can make them visible. In order to improve business results—to improve safety and resilience—you have to manage the system. Safety thinking must be system-centered—not person-centered. But, how do you know what to change—what do you need to detect and correct?

Drift and accumulation

Drift happens. Over time, the ways in which workers carry out their tasks (established practice) will wander from approved written procedures, policies, and expectations (specified practice), becoming the norm, possibly not knowing they have drifted. The more times you do something, the less concentration is expended. When people adapt to changing situations, and their adjustments succeed, and nothing goes wrong, especially more than once, it becomes the norm for that particular situation—the adjustment is perceived as a success, and overconfidence in the new "method" grows. A crescent wrench sometimes works well as a hammer, especially when a hammer is hard to come by. People begin to adopt at-risk practices. Sometimes drift is otherwise referred to by non-workers as a shortcut or a trade-off. Drift is

> the progressive or incremental deviation between expectation (*work-as-imagined*) and actual performance (*work-as-done*)—usually in the direction of greater risk (reduced margins for safety or for error).[8]

Though it's important to monitor compliance with procedures, it's more important to understand how and why the gap between expected practice and current practice grows. This is information managers can use to redesign management systems and their procedures.[9] Drift in behavior choices is more the result of system design and weaknesses than attempts to go rogue for personal reasons. This is an important perspective for managers to have when they observe work activities on the shop floor.

In her book, *The Challenger Launch Decision*, Diane Vaughan captured the essence of drift when she described the "normalization of deviance" that corrupted the ways people at NASA thought about risk and safety, which led up to the U.S. Space Shuttle Challenger disaster in 1986.[10] Because nothing bad had happened before, NASA engineers rationalized the existence of burnthrough on the inner trace of two O-rings (made of an asbestos-silica-based rubber) on the mating surfaces of the recovered segments of the solid rocket boosters—boosters that launched shuttles into space after takeoff. Initially, this was a safety concern, but over time NASA engineers convinced themselves that they understood it and it was not a concern (because nothing bad happened) and it was re-designated to a lower safety classification (less urgent). In their minds, there was no need to be concerned for the safety of the space shuttle launch system, since the outer O-ring worked. Past performance became evidence of safety, and they stopped thinking about the safety of O-rings. Failed inner O-rings became "normal."

Shortcuts and trade-offs are often adopted to accomplish assigned work—regardless of obstacles or other risky conditions. We may make a habit of making risky choices such as following another car too closely at highway speeds or choosing to drive at a high speed through a grocery store parking lot. However, do not presume that drift is all bad. All performance cannot be completely prescribed in advance. Some assumptions have to be made, such as equipment status, quality of tools, availability of resources, knowledge, skill, and readiness of workers. As unexpected situations arise, workers will be forced to adjust. The ability of people to adapt to surprises is an essential capability without which proper system functioning and safety would be nearly impossible. Hence, the need for RISK-BASED THINKING. On the other hand, workers will be tempted to take shortcuts to reduce the time and effort required. If there is a precedent for taking a shortcut, and things worked out satisfactorily, then it's a "no brainer." Do it (ironically speaking). In any case, workers tend to make trade-offs in their work being as safe as they think is necessary—particularly if (1) nothing bad happens, or (2) their bosses and co-workers say nothing to contradict it, or (3) their co-workers praise them for their "innovative" approach. Inevitably, the shortcuts become the "norm" for similar situations—i.e., drift.

Recall the near hit **event** in Chapter 3, "Big pump, little pump." Before long, Matt and Sam got used to the fact that equipment may not be labeled. So, they adapted and did their best to tag out the correct breaker. But, in many situations, these conditions exist undetected and/or they are considered unimportant, and these conditions persist, unabated—accumulation. Risks increase as defenses erode and the margin for error decreases over time. Without ongoing monitoring, deviations from safe behavior and the preponderance of faulty or missing defenses proceed unchecked—often unknown to the managers and engineers.[11] These tendencies are illustrated in Figure 7.1.

There is an important distinction between drift and accumulation. Drift is related to the gradual disconnection of people's actual practices from expectations. Accumulation is more associated with changes in the "conditions of things." Both occur gradually, over time. As unsafe conditions accumulate and the gap between

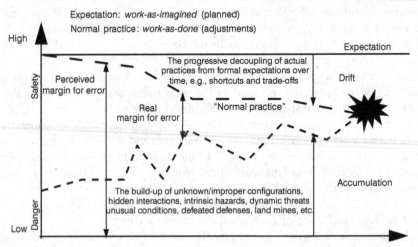

Figure 7.1 Drift and accumulation. Without a relentless management posture toward learning, at-risk practices, unsafe conditions, and faulty defenses proliferate unbeknownst to the organization's members. Eventually, everyday "simple errors" or "normal practices" trigger **events** during "normal" work. (Adapted from Figure 5.5 in Dekker, S. (2014). *The Field Guide to Understanding 'Human Error'* (3rd edn). Farnham: Ashgate.)

work-as-done and *work-as-imagined* grows, the true margin for error shrinks. Simple, everyday errors (previously trivial) possess greater likelihood of triggering harm. Lax attitudes toward **hazards** grow, and the correction of equipment issues, material defects, knowledge deficiencies, and organizational weaknesses are postponed, but safety is assumed to exist, because nothing bad happens. Tools wear out—valves, switches, software, etc. become defective after numerous uses, people's knowledge and skills decay with disuse. These conditions, inconspicuous and seemingly harmless, don't go away. They have to be identified to be corrected—discovery. Otherwise, they will build upon one another, and eventually combine with people's everyday errors or violations in unanticipated ways to trigger **events**.

As surely as a river has currents, all organizations experience drift and accumulation. Why? Because, it is impossible to know in advance what the actual work conditions will be on any given day, at any given time, including the status of key barriers and safeguards. This suggests that it is almost impossible to provide sufficiently detailed work instructions that can be followed mindlessly. Mindlessly following a procedure is commonly referred to as 'cookbooking' a procedure, reminiscent of following a recipe. The procedure user is not engaged mentally or does not have sufficient knowledge of the technology to understand what he/she is doing for each step.

Figure 7.1 suggests how accumulation leads to the incubation of your next **event**. This is another way of realizing that the causes of your next **event** exist today.[12] I have noticed over the years that managers, even when they detect "risky" situations, may "defer" corrective actions—since nothing bad has happened. They postpone corrective maintenance until later—living with an equipment issue during operation. Instead, they choose to depend on the front-line workers to "work around" (compensate for) it in the meantime. Sound familiar? Workarounds add to the operator's burden, improving the chances for error. Remember the preceding discussion about the defective O-rings on the Space Shuttle Challenger? When abnormal, risky conditions are detected—fix them now. Power plants in the U.S. nuclear power industry adopted "Fix it Now" teams—also-called FIN teams—to correct minor equipment or facility problems experienced during operations. No one has a crystal ball to predict the future, and no one knows for certain how such-and-such a condition will influence future performance. This is why a relentless, uncompromising posture toward SYSTEMS LEARNING is so important to long-term safety and resilience. *Detect and correct.*

"...safety is not a war we can fight and win. It is guerilla warfare, consistently changing and going underground."

—James Reason
Author: *Managing the Risks of Organizational Accidents* (1997)

Drift factors

There is a set of *local factors* that provoke adaptations of or breaches of formal rules and expectations and tempt people to take shortcuts. Collectively, I refer to these as "drift factors." Be alert to the fact that occasionally breaches of expectations may, in fact, be necessary to maintain safety. Don't assume a person is being reckless when they take shortcuts—it may be the right thing to do under the circumstances. Make sure you know the deeper story (explanation of the person's goals, attention, and knowledge at the time) before making conclusions about a person's motivations. In particular, operating rules and expectations contribute to drift in several ways:[13]

- Workers regularly feel pressure to trade off production requirements against strict compliance with the rules.
- Violating rules now and then is thought to benefit workers—nothing bad ever happens, and they have been successful.
- Workers, as well as managers, tend to have an incomplete understanding of safety risks and may not recognize the impact their performance can have on the system or its **assets**.
- The work itself evolves into mindless, rote performance, though risks are present.
- The work rules themselves may be impractical or simply not doable as written in the workplace.
- Too many rules may have been added—adding complexity, placing too much burden on workers.
- Workers are perceived (by managers) to protect management from liability.
- Rules, procedures, and processes are out-of-date or irrelevant to the work required.

No matter what causes drift, your job is to identify it, steering it back to safe practices.

Non-compliance with expected practices lies at the heart of shortcuts, trade-offs, and violations—and it always produces drift. From an individual's perspective the decision to drift is an economic one—does the benefit of taking the shortcut exceed its cost? The workplace can contain several conditions that tempt people to take shortcuts. The factors described below are closely related to the research done by Patrick Hudson and others on people's likelihood of "bending the rules."[14] The following relationship offers a graphical way to better understand the more relevant *local factors* that either enable or inhibit drift:[15]

$$\text{Drift} \sim \frac{B^- \ B^+ \ C \ N \ P}{S \ T \ U \ K}$$

The above factors are described as follows:

- B^- *Burden*. Mental, physical, or emotional inconveniences and encumbrances involved in performing a task; deterrents to performing an expectation according to accepted criteria.

- B^+ *Benefit.* Incentives for non-compliance with expectations such as ease, expediency, pleasure, excitement, achieving personal goals; opportunities presented to be more efficient.
- C *Control.* History of success of non-compliance, based on overconfidence in expertise, experience, or feeling of power; "I've seen it all." Nothing "bad" has happened before.
- N *Norm.* Historically accepted practice by one's peer group and/or tacit approval by those in authority; unthinking non-compliance; habit.
- P *Production.* Belief that production takes precedence over other outcomes; expectation that "breaking the rules" is acceptable to get the job done.
- S *Supervisory presence.* Bosses, peers, or even subordinates who can observe the action.
- T *Traceability.* Result of action is public and recorded such that it is traceable to the person.
- U *Uneasiness.* Sense of riskiness of non-compliance; preoccupation with avoiding harm to assets; the person's perception of personal risk.
- K *Knowledge.* Understanding the technical underpinnings of the transformation processes involved; expertise.

The factors above the bar tend to enable (encourage) drift, while those below the bar tend to inhibit (discourage) it. In all of these factors, management and leadership are the issues—many of which are the outcome of unsuitable human factors design, inadequate work planning, etc., some leading to working "on the fly." Managers, executives, and supervisors bear the responsibility for creating a system in which workers: (1) clearly understand expectations, (2) willingly and mindfully comply with procedures, and (3) understand when deviating from those procedures is necessary to assure safety. Front-line workers are given permission to make necessary adjustments in the workplace to make sure what must go right actually does go right for safety's sake only, not for production.

Healthcare hero: Ignaz Philipp Semmelweis

Ignaz Philipp Semmelweis (1818–1865) was a Hungarian physician, known today as an early pioneer of antiseptic procedures, but not so when he was alive. Described as the "savior of mothers," Semmelweis discovered that the incidence of "childbed fever" could be drastically cut by simply washing hands before entering delivery clinics. The fever was common in mid-19th-century hospitals and often fatal, with a mortality rate as high as 35 percent.

In 1847, Semmelweis proposed the practice of washing hands with chlorinated solutions, while he worked at the Vienna General Hospital's First Obstetrical Clinic. His investigation revealed that the doctors' wards suffered three times the mortality of midwives' wards. He observed that several doctors tasked with delivering babies went to the delivery clinic directly from the morgue after performing autopsies.

Despite various publications of results where hand washing reduced mortality to below 1 percent, Semmelweis' observations conflicted with established medical opinions of the time. His ideas were flatly rejected by the medical community. Some doctors were even offended at the suggestion that they should wash their hands, and Semmelweis could offer no acceptable scientific explanation for his findings. Semmelweis' practice earned widespread acceptance only several years after his death, when Louis Pasteur confirmed the germ theory, and Joseph Lister, acting on the French microbiologist's research, practiced and operated, using hygienic methods, with great success. In 1865, Semmelweis was committed to an asylum, where he died at the age 47 of pyemia, after being beaten by the guards, only 14 days after he was committed.

Today, the practice of hand hygiene has long been recognized as the most important way to reduce the transmission of pathogens in healthcare settings. But, despite this knowledge of proper hand hygiene practices, compliance is estimated to be less than 50 percent.[16]

Be wary of conformity. When people conform for the sake of group solidarity and comradery, people stop thinking, and unsafe or at-risk group norms will become acceptable and may not be recognized as unsafe.[17]

Follow through and learn

Once drift and accumulation are recognized, what can managers do about them? Dr. Patrick Hudson and other researchers have identified several strategies that can be applied, related to the drift factors described above. My own operational experience bears this out. Encouraging, valuing, and institutionalizing the following practices will go a long way toward minimizing drift and accumulation—instilling a learning culture.[18]

- *Ensure existing expectations are relevant and understood.* Is the business case for them clear—answering the *why* question? Do people understand the technical bases of procedures and instructions? Do most workers feel the rule is senseless or unwise?
- *Eliminate unnecessary expectations, policies, and rules.* Do you really need all these rules?
- *Ensure expectations are doable in the workplace.* Do they work? Are there situations where the expectations cannot be met? Have procedures and expectations been validated?
- *Reinforce and reward people for mindful adherence to rules.* Do conditions or situations exist that reward violating procedures or not meeting expectations? Is there a supervisory influence in the vicinity for high-risk work?
- *Encourage front-line workers to use initiative without taking risks.* You want people to be creative and innovative, looking for improvements not only in safety but also with efficiency and productivity. However, are people improvising for the sake of accomplishing production objectives to the detriment of the safety of **assets**? Does production take priority over safety?

Reducing drift and accumulation tends to increase the margin for error, as illustrated in Figure 7.2. Fewer errors occur, and even those that do occur have little or no impact on **assets**. Since you know drift happens, it becomes important to know how to "catch it in the act" and reverse the tendency—to nip it in the bud. Hence, the need for field observations—managers spending time on the shop floor to see, firsthand, *work-as-done* as well as weaknesses in the workplace. As weaknesses are corrected, "normal practices" will more closely and more often resemble expectations, and defenses will remain effective and robust. The workplace will contain fewer land mines. Organizational systems, values, and processes will contain fewer vulnerabilities. This is not a one-and-done effort. It's never ending. It resembles the arcade game, *Whack-a-Mole*. As in playing the game, a vigilant manager strikes every time evidence of drift and at-risk conditions raise their ugly heads. Handling issues as they arise is the primary way to avoid the accumulation of system weaknesses and unattended drift. *Don't live with problems—fix it now!*

Notice the fluctuations in drift and accumulation and its effect on the margin for error in Figure 7.2. The better the learning, the more frequent the adjustments in the workplace and in the system, and the operation becomes less reactive to people's mis-steps. This is especially true on the shop floor as front-line workers frequently detect and correct, detect and correct, detect and correct—moment by moment, day by day. Managers do the same for their organization and its systems. A robust corrective action / preventive action (CA/PA) process, for which managers are held accountable, is a highly effective way to follow through and sustain SYSTEMS LEARNING. (More information on CA/PA processes will be provided later in this chapter.)

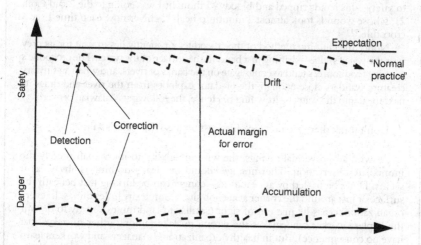

Figure 7.2 Managing drift and accumulation. As a loss of control, error represents a momentary loss of safety. By improving adherence to safe practices (better control at CRITICAL STEPS) and minimizing the presence of faulty defenses, land mines, and system weaknesses, "simple error" triggers harm less often—more resilient. Notice correction after detection

Dampening the occurrence of drift and accumulation demands that managers be relentless in detecting them daily, understanding their causes and extent, and correcting them promptly. Operations become less reactive to people's errors with increased frequency and effectiveness of learning. I believe that reducing drift tends to diminish the "frequency" of **events**, if managers provide feedback to people at the sharp end on (1) the mindful adherence to sound management expectations, (2) the exercise of good operating practices during work, and (3) the avoidance of at-risk/unsafe practices. Similarly, I believe that minimizing accumulation tends to decrease the "severity" of **events** by finding and correcting land mines and weaknesses in barriers and safeguards and their contributing system weaknesses.

Until a deficiency is discovered, understood, and corrected, the organization's **assets** endure a greater risk of harm. The remainder of this chapter is largely devoted to the discovery of system weaknesses, land mines, and other hidden flaws that tend to diminish the integrity of your system's defenses and its resilience. *Detect and correct.*

Horror on the river: nothing is always as it seems

The following episode occurred in Malawi in eastern Africa in early 2003. Dipping his arms into the murky river, Coll Sudweye (name changed to protect the individual's privacy) expected to grasp a healthy bunch of water lilies. Instead, he found both hands in the clutches of a crocodile's ravenous jaws.

The predator dragged Coll out of his shallow dug-out canoe and into the water. Memories of what followed that day in early January are sketchy. But Coll recalled the terror, the pain, and the violent struggle as six fellow fishermen beat the crocodile with their paddles.

Miraculously, they managed to free Coll from the reptile's grip and drag him to safety—his hands ripped and bloody. "I thought I was going to die," said Coll, 21, whose wounds took almost a month to heal. "I shiver now each time I see a crocodile."[19]

Notice the cunning method of the crocodile; it patiently lies in wait for its prey near the water's edge—drifting just beneath the surface of the water. The crocodile's primary food sources tend to congregate on the banks of rivers, and when an unwary creature wanders close enough, the predator explodes from the river and drags its next meal into the water—drowning it. Hence, the following Malawian proverb:

Don't think there are no crocodiles just because the water's calm.

As with all proverbial sayings, the wisdom applies to more than merely the immediate observation. The message should be clear—nothing is always as it seems. Don't be fooled by appearances—danger can be lurking just beneath the surface or just around the corner at any job site. Complexity hides things. In every organization, crocs lie just out of sight (similar to land mines), ready to devour the hapless worker who makes the wrong mistake (among other mistakes that have no consequence). But in healthy organizations, managers and workers learn where the crocs lie—they expose them. Make the hidden visible. Continually be on the lookout to know how to keep them from claiming a victim—no matter how still the waters may seem.

Learning methods

> "An accident-oriented organization waits for accidents to alert it to problems. A risk-oriented organization conducts reviews of the status quo."
>
> —Derek Viner
> Author: *Occupational Risk Control* (2015)

To comprehensively discover latent system weaknesses, land mines, faulty defenses, and error traps in complex systems, a diversity of methods and inputs into problem detection are necessary. Discovery is followed by understanding from a systems perspective, which leads to more effective, sustainable corrective actions. Because of the variety, extent, and, in some cases, the faintness of unfavorable conditions in organizations and their facilities, you cannot rely on any one method to detect all the issues. Although not a complete list, the following learning methods have historically proven themselves effective in identifying and resolving system problems:

- reporting
- field observation
- learning teams
- metrics and trending
- self-assessment
- operating experience
- benchmarking
- **event** analyses:
  - local rationality analysis
  - local factor (gap) analysis
  - behavior choice assessment
- corrective action / preventive action process.

There are no new big "ah-ha's" on this list—various versions of these methods already exist in most enterprises. But, too often, SYSTEMS LEARNING is given lip service relative to the priority given to production. Therefore, it takes commitment and accountability by executives and line managers to discover the "causes of your next **event**" so that corrections can be made before suffering serious harm to your organization's key **assets**. The effectiveness of these learning methods is a function of management's general posture toward "bad news"—especially information that has the potential to be politically embarrassing. Senior managers must possess the courage to do the right thing that improves the safety and resilience of the organization and its workforce, even when it looks bad.

Important. Learning requires humility—without it, you will not learn. When you think you know it all, you'll stop asking questions or soliciting feedback. Do not take offense when your expectations and systems don't work as you thought they would. Most strategic decisions must be modified during the process of implementation.[20] Your willingness to admit mistakes encourages others to do the same. In the long term, humility builds trust.

Data from these feedback sources are regularly reviewed to identify issues for heightened management attention. This information gives management a current and objective picture of system safety and resilience capabilities and limitations. In the end, detection has no value if problems are not corrected. Because of the ever-present lure of production demands, managers must be accountable for following through to improve their organization and management systems concurrent with operations. *Safety cannot be isolated from production (business) processes or vice versa!*[21]

Reporting

Just as the wheels of a car must be in alignment—or else the car will drift off the road—an organization needs to have its working parts aligned in order to stay on track to achieve its desired business results. Managers need to know about surprises and noteworthy and recurring differences between what they expect to happen and what actually happens, especially for high-risk operations. Because a surprise is a warning signal, any surprise—no matter how inconsequential it may seem—should be communicated quickly. If an organization does not take surprises seriously, "minor" occurrences become normal—and over time they go unreported. The lack of reporting itself becomes a latent system weakness.

Feedback from front-line workers is your primary source of real-time insights into the effectiveness and vulnerabilities of the organization's work-management system. Soliciting their feedback and using their input are crucial to sustaining system alignment. I address reporting first among all the other learning methods because it taps into the richest and most valid source of information about the functioning of your system. "Deference to expertise"[22] is a key success factor of high-reliability organizations. Front-line workers, who do the work, are the obvious specialists in their work.

"Managers, ask for information you need to know, not what you want to hear. Workers, tell them what they need to hear, not what you want to tell them."

—Roger Boisjoly
Former chief engineer, Morton-Thiokol, Inc. (1998)

Systems thinking managers actively promote an open reporting culture—they do so on the assumption that what seems to be an isolated failure or error is more likely to be the confluence of many upstream factors.[23] Therefore, front-line workers must understand the importance of their reporting—that it is viewed as essential to better understanding the true state of the system. Let me make it very clear that one of "the most harmful of harmful practices is the failure to report unsafe conditions, behaviors, actions, and inactions, including purposeful non-reporting."[24] You must ensure front-line workers understand, in no uncertain terms, that they have an obligation to raise the red flag when they see something that is not right, unusual, or surprising. Just make sure you do something with the information, and let them know what you did about it.

An open reporting culture requires trust. It gives people at all levels the freedom to report issues—without fear of reprisal or punishment or assigning blame to individuals.[25] Front-line workers must perceive that their managers are fair, equitable, and willing to act rightly to the facts, rather than punishing people for admitting to a mistake. Otherwise, they will ask the following questions:

- Will my manager do something to improve the situation?
- Is it worth the extra effort and risk to me personally, when no real good is likely to come from it?
- Will I get my co-workers in trouble?
- Will I get in trouble? Are mistakes seen as failures?

Dr. Reason calls this a reporting culture, which he asserts must be engineered.[26] That is, the organization is aligned via the *local factors* to encourage reporting.

"Managers who don't understand the operator's point of view at the sharp end can miss the demands and constraints operators face."

—Christopher Nemeth
Author: *Resilience Engineering in Practice,
Volume 2: Becoming Resilient* (2014)

Reporting works best when it's fast, easy, and simple to do. An effective reporting system has two key characteristics: (1) it maximizes the ease of reporting, and (2) it minimizes anxiety about reporting.[27] As emphasized already, to reduce anxiety about reporting, managers must establish trust that the information provided in good faith will not be used against the one who reported it—a "just culture."[28] Regardless what you call it, trust is essential for all forms of personal disclosure in an organization—but it's hard to build and easy to lose—and line managers must be able to cultivate and nurture it. At this point I want to postpone the discussion on accountability and just culture. It's important that managers understand it and how to influence it. Chapter 8 provides greater insights and principles about building a just culture in the context of accountability.[29]

Field observation

Most problems in organizations exist because managers are unaware of them. The best way to combat this is simply to go watch what happens in the workplace. If you truly want to understand what is really going on, walk into the production spaces of your operation and see, firsthand, what people are doing, what conditions they are working in, what they are working with, and how they are coping with less-than-perfect conditions to accomplish their assignments. Work is the only place in a business where value is added. This is the essence of observation.[30]

An observation involves close monitoring of the performance of a person's or team's work activities to (1) see firsthand how work is performed, (2) understand its context (*local factors*), and (3) assess the results (are you getting what you want?). The observer focuses on both the worker and supervisor behaviors, as well as on the conditions affecting (1) their behavior choices, (2) the safety of **assets**, (3) compliance with regulations and quality requirements, (4) productivity, and (5) efficiency. Some people refer to observation as an "engagement." Managers have to engage the workforce. Engagement doesn't happen unless learning occurs. At least one person, the worker or the leader, has to learn. Preferably, both learn.

First-hand observation can be labor intensive and expensive, requiring the careful selection and training of those who observe. Another disadvantage of field observations is that they can influence the behavior of those who know they are being observed. This is overcome by avoiding "drive by" observations, spending no less than 45 to 60 minutes watching an activity. Decide (1) what to observe (preferably people engaged in high-risk activities), (2) who will conduct the observations, and (3) when, where, and how often to observe. This learning method's effectiveness depends on the observer's comprehension of the **Hu Risk Management Model** and the control of **pathways** and **touchpoints**.

Observing work in addition to all the other responsibilities on your plate of responsibilities can be overwhelming. Do you really need this? Is it worth your time? The reality is if you're in a leadership position, people look to you for direction, and you are a coach by default. Your presence on the shop floor signals to the workforce the importance you place on the conduct of operations— encouraging high-levels of performance and adherence to expectations. Chapter 8 provides more information on the conduct of observations.

Caution! If an incident is the result of an individual taking a short cut, it is unlikely that it occurred the first time the short cut was taken. It is more likely that short-cutting has been going on for months, perhaps years. A good manager would have spotted it and stopped [corrected] it. If he does not, then when the incident occurs, he shares the responsibility for it, legally and morally, even though he is not on site at the time.[31]

Finally, I must leave you with one very important admonition. It is important to identify and minimize drift from expectations. But, you must do so in ways that sustain people's dignity. In all your interactions with people, the overall attitude of leaders in positions of responsibility should be to believe the best, want the best, and expect the best of people. People want to be effective and to be successful. Unless there is strong evidence to the contrary, people want to do a good job. Always treat people with respect, honesty, and fairness.

Learning teams

As pointed out in the section on *Reporting*, who knows better what works and what does not work than the front-line worker? It is worth saying again—the front-line worker is the richest source of information about the functioning of the organization's systems and processes. Dr. Todd Conklin, during his time at Los Alamos National Laboratory, devised a means to tap into this source of information, which is described in detail in his book, *Pre-Accident Investigations*.[32]

Peer groups consisting of workers, supervisors, and technical specialists can be brought together on an ad hoc basis to assess **events**, *work-as-imagined* vs. *work-as-done*, near hits, and, even, work-arounds—anything the organization wants to learn about. A learning team's primary purpose is to learn and communicate **event** information to management. When facilitated properly, learning teams create a forum that encourages informal and open discussions among team members on a specific issue. When used proactively, learning teams surface crucial safety concerns before they become a serious problem. Learning teams can be formed for any of the following occasions:

- pre-critique as part of a pre-job briefing (before performing risky operations);
- post-**event** (**event** analysis);
- near hits or close calls (**event** analysis);
- interesting successes (*work-as-done* vs. *work-as-imagined*);
- anytime you can't explain unexpected occurences or conditions.

In his book, Dr. Conklin suggests that learning teams can be formed at a moment's notice—ad hoc, or they can exist permanently for particularly important functions that must go right every time. Effective learning teams:

- limit their size to 4–5 members (no one can hide; easier to reach consensus);
- start with free-flowing dialogue about the issue (a sort of brainstorming);
- seek a balance of interpersonal skills (enhances diversity of perspectives);
- sink or swim together (promotes interdependence—"we're in this together");
- challenge each other (probe assumptions in search for facts);
- test the benefit of the change (validate recommendations).

Learning team members are directed to explore "how" something happened, not "why" or "who." "Why" tends to focus people's thoughts on the structural cause-and-effect features of the issue. Talking about "who" fosters defensiveness—inhibiting honest disclosure of facts. The only real structure the learning team is given is that they are to tell a clear story about "how" something happened (or could happen):

1   *beginning*—context of what was happening
2   *middle*—consequences of mishap (or potential consequences)
3   *ending*—understanding of how **event** occurred or how the situation is risky.

The team prepares a simple, written report about what the organization should learn from the issue, followed with a verbal briefing with the appropriate level of management.

Notice that the structure is quite informal, which is by design. The intent is to facilitate the flow of insights about an issue and pass them on to managers—not to create an auditable document or a paper trail. Resist the temptation to formalize the process. As the learning-team process becomes more formalized, you will lose honesty. Make the tool work for you; don't make the workers work for the tool.

Metrics and trending

An indicator relevant to H&OP is a measure that provides information on (1) the occurrence of human actions (behaviors, errors, violations, etc.), (2) the existence of various conditions (defenses, values, equipment status, etc.), or (3) the direction of key result areas (safety, production, quality, etc.)—the slope of a value over time.

"I like good news as much as the next person, but it also puts me in a skeptical frame of mind. I wonder what bad news I'm not hearing."
—Bill Gates
Author: *Business at the Speed of Thought* (1999)

Managers should be on the lookout for patterns, especially how defenses (controls, barriers, and safeguards) fare over a period of time.[33] Trending involves the analysis of various sources of performance data that leads to the recognition and correction of system-related weaknesses and problems. Trending makes the invisible visible. Over time, an array of issues, when reviewed in aggregate, will reveal systemic weaknesses, not recognized previously, that contribute to drift, weak or missing defenses, and unsafe workplace conditions such as land mines. Trending supports the recognition of systemic issues by providing a means to statistically analyze data and metrics collected during operations. Once recognized, an organization should be able to repair a problem before it reveals itself through an **event** or other significant organizational breakdown. It is not my purpose here to review measurement science and practices or explain how to do trending. My intention is to emphasize its importance and several key factors that influence its effectiveness.

The question is "What do you want to know?" How do you know your operation is safe? One way to know is by managers meeting collegially on a regular basis to review indicator trends for the following purposes:

- to identify threats to the safety of **assets** and to resilience;
- to identify statistically significant declining performance in key mission areas;
- to assign accountability for developing and implementing corrective actions;
- to monitor the effectiveness of corrective actions.

Warning! The goal is not simply to collect data. The primary purpose of trending is to preserve the safety of the organization and its **assets** and to sustain its resilience. Ultimately, the data is to help people by making changes to the system that make desirable outcomes more likely and unwanted outcomes less likely—it's not to judge people.[34]

Managers tend to rely on result-based metrics—lagging indicators, which are after the fact and tend toward zero. A lack of indication of harm (such as a low count of **events**—a lagging indicator) is often assumed as evidence of safety. If all lagging indicators were zero, how would you otherwise know safety is present? Through leading indicators. Recall that **Hu** is a combination of behavior and results (**Hu** = B + R). Lagging indicators are derived from results, while leading indicators provide information about behavior. Preferably, you want an indicator that increases as safety is enhanced.

One central principle of H&OP is "Safety is what you do." Safety is confirmed by the presence of something that creates safety—defenses (controls, barriers, and safeguards)—resilience. H&OP is about controlling, learning, and adapting—preserving the safety of **assets** and optimizing the resistance of systems to **events**. To really know if something is safe, you must select and monitor measures that confirm its presence—measures of what is happening or currently exists—leading indicators. They are leading if they provide information such that actions can be taken in time to prevent an unacceptable change in one or more key result areas (such as safety, production, quality, customer satisfaction, and economics). Leading indicators reflect the resilience of an organization—the real-time measurement of current practices and conditions, such as the occurrence of pre-job briefings for high-risk work, stopping work when unsure, the amount of time defenses (barriers and safeguards) are circumvented or out of service, the number of reports submitted after work, training completed, observations conducted, high-risk jobs with CRITICAL STEPS identified before starting work, number and age of workarounds, and corrective actions completed on schedule.

Warning! Generally, it is not worthwhile to count errors. By acknowledging the fallibility of all human beings—that **Hu** is never completely error-free, the need for an indicator of "human error" disappears. However, it may be worthwhile to identify which activities and equipment experience losses of control.

What does a leading indicator look like? Dr. Erik Hollnagel, in his book *Safety-I and Safety-II*, suggests using the generic *n/m* indicator to develop measures to monitor the occurrence of desired behaviors or conditions important for safety, which increase as things get better.[35] The *n/m* measure is a generic leading indicator, where *n* represents the number of positive occurrences, which is divided by the total number of occurrences, *m*. For example, you may want to know the portion of high-risk jobs that receive a pre-job briefing. You could measure the number of work activities with at least one CRITICAL STEP that receive a pre-job briefing compared with all work activities with at least one CRITICAL STEP. To effectively monitor **Hu**, you must measure behaviors and practices (leading) as well as the results (lagging).

After recognizing an adverse trend, managers must determine why those deficiencies exist and how widespread they are, and identify corrective actions to remedy them. But, do not be so quick to respond. A common management error is taking premature action on an apparently "undesirable" trend, when, in fact, there is no adverse trend—only normal statistical variation.[36] Also, executives frequently use such indicators as justification to redistribute limited resources away from safety.[37] Just because the "metrics" look good does not necessarily mean you're safe right now.

To sustain accountability, management reviews of metrics should occur no less often than every 90 days or quarterly, depending on the tempo of operations. In some high-risk, high-tempo operations, managers meet weekly, or as needed, to review key indicators.

"Making regular checks on the 'vital signs' of an organization is what management is paid to do: it is not safety specific, it's just good business."[38]
—James Reason

How do managers and executives apply a chronic uneasiness toward their metrics? First, managers must accept red indicators for what they mean—believe the worst until actual performance is otherwise validated. They must resist the temptation to rationalize a deteriorating indicator—to explain it away or minimize its significance. Second, managers should challenge green indicators—treat all positive results with skepticism until verified independently. Also, they should not postpone appropriate improvements for the sake of managing the "optics" (political posturing) of indicators.

There is no single reliable measure of safety and resilience. Instead, managers need a blend of both lagging and leading indicators to help them manage H&OP. It is necessary to know what should be happening to verify the presence of safety. In contrast, it's also vital to know when safety and resilience wane—foreshadowing unwanted outcomes.

Warning! Do not use metrics related to individual injury as an indicator of system safety. Individual accidents and organizational accidents have different causal sets.[39]

The selection of which indicators to use depends on management's understanding of their organization, as well as how safety and resilience are created in their workplace. If a manager wants to change x, he or she should start by measuring and managing the drivers of x.[40] But, to help managers identify such drivers, the selection of safety indicators should be soundly based on an underlying, but validated, model of safety.[41] To this end, the *Systems Thinking Model* and the Building Blocks of Managing H&OP provide a couple of frameworks to guide managers in monitoring trends in human and organizational performance. Appendix 4 describes candidate leading and lagging indicators useful in managing H&OP.

Self-assessment

Over time, legal requirements, technology, and industry best practices change. Drift and accumulation happen. What was an acceptable level of risk control at the time of commissioning is likely no longer acceptable. There is a need for periodic review of processes, programs, and expectations—self-assessments.[42]

Self-assessment uncovers problems by comparing actual performance (behaviors and conditions) to currently recognized best practices. In addition to practices in the workplace, a key focus of assessments is the effectiveness of defenses in high-risk operations and related safety systems: their robustness, reliability, availability, ease of use, delay in implementation, dependency on humans, and their costs.[43] Analogous to an audit in accounting, a self-assessment is likely the most effective means to discover latent system weaknesses. Some forms of self-assessments are on-going, such as field observation, supervision, and quality control. Others are formally performed on a periodic basis and when needed, such as an **event** analysis.

Periodic, formal self-assessments identify safety concerns and related system weaknesses. Managers deliberately discover for themselves the flaws in their defenses—not only at an operational level but also within management systems, production processes, facilities, culture, leadership practices, training programs, and the organization in general. This learning method offers significant protection against the accumulation of latent system weaknesses and the buildup of hidden **hazards** in the workplace.

Using best practices, an organization can identify gaps in performance, processes, and methods by comparing current performance, processes, practices, and standards against those standards. Corrective actions can then be instituted for specific issues. Recognizing patterns (trends) in which defenses fail—such as recurring equipment failures, error traps with specific error types, and lack of procedure use—provides opportunity to improve processes, programs, and policies. A policy and program to establish and guide the conduct of self-assessments should include the following elements:

- Ongoing and periodic self-assessments used to identify safety concerns and to improve performance by comparing current performance to best practices.
- The self-assessment process applies to all levels and functions of the organization—especially those important to safe operation.
- Self-assessments compare actual performance to (1) current management expectations, (2) performance of other similar, high-performing organizations (see benchmarking), (3) requirements of industry oversight organizations, and (4) regulatory requirements.
- Improvement needs identified by self-assessments are assigned for action and tracked through completion (see section on *corrective action / preventive action (CA/PA) process*).
- Skilled, knowledgeable employees and qualified outside personnel perform self-assessments.
- Self-assessment results are communicated to the affected work groups and individuals.
- The effectiveness of the self-assessment process is periodically assessed—and adjusted—based on the results.
- Indicators of the self-assessment process effectiveness are developed.
- The chief executive officer or appropriate senior manager monitors the effectiveness of self-assessments and independent assessments.

Experience has shown that establishing and implementing the self-assessment process are not sufficient steps—by themselves—to ensure its effectiveness. Success depends on regular accountability among managers and follow-through with corrective actions. Though the scheduling, resourcing, and conduct of self-assessments—and the implementation of related corrective actions—is a slow process, managers can neither take shortcuts nor make trade-offs for the sake of efficiency and cost reductions. Executive commitment must be present if self-assessments are to be successful in identifying and resolving latent system weaknesses that influence safety and resilience. For this reason, it would be wise to periodically conduct an "independent assessment" of the self-assessment process to get an unbiased opinion on its effectiveness.

Operating experience

An old adage says, "Anyone can learn from his own mistakes; a wise man learns from the mistakes of others." The effective use of operating experience (OE) gives an organization the opportunity to make wise use of the mistakes others have made. The intent of an OE management system is to prevent or mitigate the consequences of similar **events** based on the experiences of others. Sometimes lessons learned are based on **events** from within your own organization, but most of the time they are from **events** from outside organizations. Either way, when a company reviews OE—and disseminates information about positive and negative experiences relevant to operational tasks—employees throughout the organization become more aware of vulnerabilities.

It is common to monitor occurrences from organizations with similar technologies, such as biotech facilities learning from the **events** of other biotech facilities. However, many organizations look alike regardless of their technology. The lessons that could be learned should not be restricted to operational levels, or to organizations and facilities within your industry. Look outside your industry.

Effective use of OE requires learning. The information from the **events** of others is often referred to as lessons learned. Lessons generally refer to the corrective actions to address the loss, damage, or injury the suffering organization experienced, and the preventative actions that can be done to avoid similar consequences or to secure desired outcomes. Lessons may reveal vulnerabilities with in-house operational methods, flaws in technical processes, and weaknesses in management systems. OE is particularly useful just before commencing work. During pre-job briefings, front-line workers review **events** related to the work they are about to perform, asking themselves what they will do today to avoid the mistakes and consequences of the **event** in question.

OE can re-direct the tendency of people to think "it can't happen here" or "that won't happen to me" by showing that it can, indeed, happen anywhere. A wise manager incorporates lessons learned into organization and management systems, applies them to operational processes, and relentlessly underscores applicable lessons in every relevant forum: training venues, pre-job briefings, meetings, coaching by supervisors, engineering design reviews, and even executive decision-making at a strategy or policy level.

Based on experience in the military and the nuclear and aviation industries, the effective use of OE is optimized by the following success factors:

- Institutionalize a management system process for the use of OE—make it mandatory.
- Ensure that lessons learned are translated into corrective and preventative actions that improve safety, resilience, quality, efficiency, etc.
- Make OE information easily accessible to facility personnel, especially those at the sharp end, for use in work planning, preparation, and training activities.
- Distribute relevant information to operational personnel in a timely manner.
- Deliver OE information in a story format using active learning methods.

- Share in-house lessons learned related to internal **events** with outside organizations.
- Conduct a self-assessment of OE management systems every two to three years. (I recommend annually for new or immature processes.)

To internalize the lessons learned from OE, people must ask themselves how they would personally avoid (1) similar mistakes in similar circumstances, and (2) similar consequences to the **event(s)** in the OE report.

Benchmarking

As OE learns from the mistakes of others, benchmarking alternatively promotes learning from the successes of others in a particular domain of performance. It does so by comparing the organization's current practices, such as operations, with those of high-performing organizations. Benchmarking usually involves sending individuals or teams to a specific organization or to industry conferences, where specific topics of interest are discussed. This provides a way to identify internal vulnerabilities, performance gaps, and improvement opportunities.

By comparing an organization's practices with the patterns of others considered "best in class," the organization can see more clearly where it has room for improvement. Benchmarking highlights relevant differences between best practice and internal practice. This helps shed light on internal weaknesses—areas needing improvements. Additionally, benchmarking helps line managers remain aware of high standards of performance—i.e., industry "best practices." However, managers too often use benchmarking as an excuse that "we are as good as they are" and, therefore, good enough. This is a *big* problem in healthcare systems. Too often, I've seen managers, not committed to systems learning, go through the motions of "improving," but doing nothing to improve safety.

To make the best use of benchmarking, managers must:

- elicit the assistance of senior front-line workers who possess high levels of expertise (subject matter experts);
- gather information systematically on pertinent performance issues;
- understand best practices;
- document improvement targets;
- assign responsibilities;
- monitor progress through corrective action;
- avoid "industrial tourism"—visiting other facilities, but not identifying or following through on improvements within their own organization.

Benchmarking activities have a clear scope, set of objectives, and deliverables. Opportunities as well as weaknesses and vulnerabilities identified through benchmarking are assigned for action and tracked through completion, as are other corrective actions. Always consider organizations from other industries for benchmarking.

Personal experience: navigating a submarine up river

As a line officer deployed on a U.S. nuclear submarine in the late 1970s, I once had the opportunity to navigate the Cooper River from the Atlantic Ocean to the Charleston Naval Base several miles inland in South Carolina.

To avoid running aground—causing serious damage to the submarine's hull and equipment—we needed to stay in deep water, carefully avoiding shoals. To know exactly where we were in the shipping channel at all times, the ship's navigator (with the aid of several others) recorded bearings to fixed landmarks along the river and plotted them on a navigation chart (map) of the river channel. To optimize his confidence in the ship's position on the chart, the navigator always asked for at least three bearings (lines of position)[44] from three dispersed landmarks—not two—to more reliably pinpoint the ship's location in the river. Sometimes, due to inaccuracies with instruments, variations in taking the bearings, and in drawing the lines on the chart, the chart would reveal a small triangle. Regardless, there was a higher degree of confidence that the submarine was somewhere within the triangle than if only two lines of position were drawn (a single point). Obviously, one line of position is insufficient to create a point on a chart.

In a similar way, to optimize the reliability of the conclusions reached about a person's actions or inactions that triggered an **event**, I recommend analyzing the **Hu** in the **event**, using three slightly divergent perspectives—lines of inquiry. This requires the analyst to think about the occurrence from different perspectives. That way, there will be greater confidence in the revealed "causes" of an **event** and in its corrective actions to reduce the chances of recurrence of the same or similar **events**.

Event analysis

If proactive learning methods have been ineffective, eventually the organization will learn about its weaknesses through **events**—learning late. **Event** analysis is a discovery process designed to analyze and understand the systemic reasons for **events**. To improve the integrity and reliability of the analyses of **Hu events**, I suggest your analysts look at each **event** using three lines of inquiry to confirm their causes:

1 *Local rationality analysis*—to understand people's rationale for their choices
2 *Local factor analysis*—to identify the *local factors* contributing to the difference between desired and actual behaviors
3 *Behavior choice assessment*—to rule out sociopathic behavior and to avoid hasty reactions by managers to punish "honest mistakes."

The following sections briefly describe these three analytical approaches to understanding **Hu** from a systemic perspective. But before we do, it's important for analysts and their managers to understand that how they think about individual performance and how organizations work significantly influence how the analysis proceeds. All analysts have mental models, explicit or implicit, that guide their thinking during analysis.

Events are fundamental surprises. The causes of **events** permit the chance combination of **pathways** between intrinsic **hazards** and **assets** and related human actions, involving the concurrent breach of all the barriers and safeguards, to cause harm.[45] As fundamental surprises, **events** are organizational failures—managers did not fully understand how their organizations work and were surprised by the outcome. It should then be self-evident that **event** analysis is about organizational learning, not placing blame. Answering the following two questions will guide managers to what needs to be corrected organizationally, systemically.

1 How did it happen?
2 Why were you surprised?

Both questions seek insights about the inner workings of your system. The first question attempts to expose the real, deeper story of how normally conscientious people ended up doing what they did, while the second question seeks to explain management's lack of understanding about how their system worked—how defenses either failed to protect **assets**, were circumvented, or were not even present.[46] Collectively, both questions look at how RISK-BASED THINKING didn't work, individually and organizationally.

Notice the verb "how" in the first question. Understanding how errors or choices were made helps you understand the local conditions that influenced them, which point to likely system weaknesses. To explain failure, do not try to find where people went wrong. Instead, investigate how people's understanding of the situation and actions made sense to them—at the time, given the circumstances that surrounded them.[47] Analysis focuses more on discovering the manner in which a normally "safe" system failed to protect its **assets** from harm, not why—not "who screwed up." Corrective actions, from a why perspective, are usually oriented toward repairing or replacing the "failed components" without understanding how they interacted or emerged to spawn the consequences. Answering "How?" explains how people lost control of **hazards**, **pathways**, and their actions during operations. Understanding surprise helps explain how defenses failed.

Human error is *never* a root cause!

Recalling the *Systems Thinking Model* (see Figure 6.1), you should realize that human error is not the stopping point of **event** analysis—rather, it's the starting point. Too often, once human error is discovered, analysis tends to stop prematurely without fully knowing the back story. The analyst senses that the defective component has been identified—*the* root cause—and simply needs to be corrected or "changed out." Such a conclusion is incorrect. Human error is a "red herring"—evidence of something deeper.[48] To avoid reaching such a conclusion, an **event** analysis should clearly show the causal relationships (avenues/linkages) from consequences to the relevant organizational and system weaknesses—not just the individual!

Figure 7.3 illustrates the *Systems Thinking Model* from a failure perspective—the *Systems Failure Model*.[49] As you can see, errors that occur at the sharp end of the organization, occasionally trigger a loss of control during operations. As illustrated, an active error—an error with immediate adverse consequences for **assets**—has its roots in the work context—the conditions people work in (*local factors* / error traps). These conditions, in turn, have their origins in the form of various latent system weaknesses. Similarly, defenses (controls, barriers, and safeguards) are circumvented or otherwise become ineffective in protecting **assets** from harm through various undetected system avenues. As you can see, **event** analysis has a bottom-up route—sharp end to blunt end. The *Systems Failure Model* helps you better understand how an **event** occurred.

Harm and its severity always involves failures of defenses in some form or fashion. In every **event**, defenses (controls, barriers, and safeguards) either break down or are circumvented to allow the uncontrolled transfer of energy, movement of mass, or transmission of information in a way to produce harm. Describing the form and mechanisms of harm to **assets** and the breakdown of defenses suggests how the organization failed to protect **assets** from harm. Organizational weaknesses are manifested in the workplace via the latent system avenues (organizational pathways) that defeated or circumvented defenses or created land mines. The *Systems Failure Model* helps the analyst identify the true causes of any serious **event**—the blunt end—the organization and its management systems. This helps you understand why you were surprised.

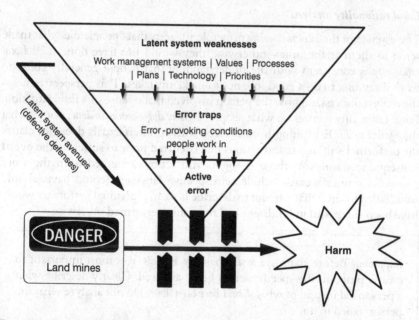

Figure 7.3 The *Systems Failure Model* illustrates what the *Systems Thinking Model* would look like in an **event**. The organization's role in **events** is clearly evident via various system avenues. (Adapted from Figure 1.6 of Reason, J. (1997). *Managing the Risks of Organizational Accidents*. Aldershot: Ashgate.)

Analysts must discipline themselves to "see" the system and its avenues during the analysis, working backward from the **event** consequences through workplace behaviors and the *local factors* to the organizational weaknesses that "caused" the **event** and similarly for faulty defenses. Recall the *Alignment Model* in Figure 3.2. This is the "how." It helps explain how a person did what he or she *did* at a particular point in time (not what they failed to do) and how defenses failed allowing harm to **assets**. Using the *Systems Failure Model* along with the *Alignment Model*, the analyst should explore the following during the analysis:

1   Harm to **assets**—the form, magnitude, and intensity of consequences.
2   The **pathway** between the intrinsic **hazard** and **asset.**
3   The CRITICAL STEP that released the **hazard.**
4   Characterize the human initiating action (adjustments, performance gaps, active errors, or violations) at the CRITICAL STEP.
5   *Local factors*, error traps, or drift factors that prompted the initiating action.
6   Missing and ineffective defenses that failed to either prevent the initiating action, or prevent or mitigate harm.
7   The latent system weaknesses (*organizational factors*) that contributed to items 5 and 6.[50]

Let's look at each line of inquiry in turn.

Local rationality analysis

The essence of the *local rationality* principle suggests that "people did what made sense to them at the time—otherwise, they would not have done it."[51] *Local rationality* is consistent with the understanding that people generally come to work intending to do a good job, not to make mistakes. This perspective helps the analyst understand how the person involved made sense of a fluid situation. Local rationality focuses on what people actually *did—work-as-done*, not on what they *failed to do*. Explaining human error with counterfactuals does not clarify the performer's thoughts and actions that occurred prior to suffering the **event** consequences and how those thoughts and actions emerged from the work context. Counterfactuals include retrospective language (could have, should have, failed to, etc.) that attempt to describe how an individual performer would have been successful in hindsight had they only performed otherwise.

> *Important!* Before starting a *local rationality analysis*, it is most important to clearly pinpoint the specific action being assessed. Clearly describe what a person *did* instead of what *should have been done*. Do not analyze what the person failed to do.

Work must be understood from the perspectives of those who did the work—their reasoning for the behavior choices made at a particular time. This approach helps identify the person's immediate rationale for their decisions through understanding his/her goals, foci of attention, and knowledge of processes at hand.[52] To better understand one's reasoning and perception of the situation, the analyst can ask the following questions—notice there is some overlap in scope:[53]

- What was he/she trying to accomplish? What other goals existed at the time?
- Where was his/her attention focused? What did he/she see/hear/feel at the time? What was he/she trying to control?
- What was his/her mental picture and understanding of the technical situation as it was developing?
- What always happens—normal? What never happened before—unusual?
- What is the relevance of the difference between what happens all the time and what had never happened before?
- What was rational—in the here and now—with respect to the streams of activities? How were critical parameters trending?
- What decisions/adjustments/mistakes were made? What was expected?

Time is an important factor in understanding any **event**. I am a strong proponent of charting events and causal factors. Instead of describing charting here, I suggest you explore this sequence diagraming technique as described in William Johnson's seminal book, *MORT Safety Assurance Systems*.[54]

The analyst attempts to explain why it made "perfectly good sense" to an individual performer to do what was done instead of what was supposed to happen—a performance gap. The assumption is that people do not intend to err or to fail. A "performance gap" is a statement of what happened instead of what was supposed to happen. Using Kelly as our example, "Kelly proceeded with the task when unsure instead of stopping to get help." (See the case study, "First impressions are last impressions," in Chapter 4.) A performance gap is characterized by filling in the blanks in the following:

While performing _____ (task, action, operation),
the individual _____ (actual behavior—*work-as-done*),
instead of _____ (desired behavior—*work-as-imagined*)
resulting in _____ (outcomes, consequences).

The next line of inquiry attempts to understand the factors surrounding the person that led the individual to draw the "wrong" conclusion about what to do. Clarifying the performance gap as illustrated in Figure 7.4, helps the analyst better understand why it made sense to do what was done and not do what was expected, which is the objective of the next line of inquiry.

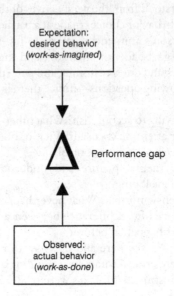

Figure 7.4 The performance gap

> *Caution!* Be careful not to presume that *work-as-done* was wrong. What actually happened may have been appropriate. Perhaps, *work-as-imagined*, the expected behavior, may have been inappropriate for the situation.

Using the proposed performance gap statement, it is validated by the individual(s) involved and corroborated with information relevant to the **event**. The following "test for validity" can be applied to ascertain if the person's action did, indeed, trigger the consequences. It is a straightforward logic test. The question is:

> *Validation test.* Would the **event** (consequences) have occurred if the individual's action (the performance gap) had not occurred?

If the answer is no, you can be confident that the individual's action initiated the accident. The credibility of the analysis is enhanced if two or more other individuals answer this question independently.

Local factor analysis

Now that the performance gap is clearly defined and validated, the analyst can conduct a local factors analysis (see Figure 7.5). The intent of local factors analysis is to explore the context of a person's choices and action by identifying specific and relevant workplace conditions that:

- enabled or inhibited the *actual* behavior (*work-as-done*);
- enabled or inhibited the *expected* behavior (*work-as-imagined*).

Using Appendices 1 and 2 for this analysis, this line of inquiry helps keep the analyst focused on the system and not on the superficial failings of the individual. Each relevant *local factor* points to one or more respective *organizational factors*— avenues of influence—that fundamentally provoked the **event** (root cause).

The performance gap suggests there were *local factors* present in the workplace, either working to enable the actual behavior or insufficient drivers to influence the expected behavior. Alternatively, there may have been *local factors* that either inhibited the expected behavior and/or failed to inhibit actual behavior. As you can see in Figure 7.5, some *local factors* are stronger or weaker than others. The simultaneous combination of all these factors influenced people's choice at the time. The outcome of this analysis is compared with the local rationality analysis performed earlier, which should corroborate the local factor analysis. Once these various factors are validated, the analyst now has more questions to explore related to the organizational and systemic factors contributing to the particular *local factors* influencing the choices made.[55]

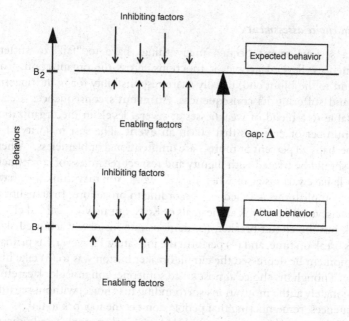

Figure 7.5 Local factor analysis. The illustration explains the gap between actual behavior (*work-as-done*) and expected behavior (*work-as-imagined*). In each case, there are local factors that enable and inhibit the behavior. (Adapted from Figure 6.3 in Chevalier, R. (2007). *A Manager's Guide to Improving Workplace Performance*. New York: AMACOM.)

A more detailed discussion of *local factor analysis* is provided in Chapter 8 (see Figures 8.2 and 8.3). As before, to help validate the strength and relevance of each factor, the analyst can ask the question:

> *Validation test.* Would the individual's action (*work-as-done*) have occurred if the particular local factor had not been present?

If the answer is positively no, then there is high confidence in the relevance of the *local factor*. A similar logic test could be made for *local factors* related to expected behavior.

The behavior choices of front-line workers are a reflection of how well their managers are doing their jobs of managing their systems.[56] People do what is reinforced, and there are organizational or systemic reasons for these conditions. *Remember, whatever people are doing pinpoints the reinforcers.* A reinforcer is a *local factor* (cell 3 (of Table 3.2 and Appendix 1)) that incentivizes or strongly encourages a specific behavior, whether desired or undesired.

The third line of inquiry focuses on a person's intent. It too assesses a person's choices in light of the context. I refer to this method of adjudication as a *behavior choice assessment*.

Behavior choice assessment

When a serious **event** occurs, many things have to "fail" to suffer such consequences. Several people or functions across the organization, from the sharp end to the blunt end, usually share responsibility for both triggering the **event** and suffering its consequences. An **event's** consequences are usually attributable to a blend of weaknesses at several levels in the organization due to the number of defenses that fail in an **event**. The vast majority of unsafe acts—perhaps 90 percent or more—are unintentional or blameless.[57] Therefore, people should be treated with dignity and respect regardless of consequences.

People have various goals when they do work. You may think they have only one—accomplish the assigned task according to procedure. In actuality, people may have several. Let's look at the goals of Kelly, our newly qualified electrician: (1) rack out the circuit breaker, (2) demonstrate his newly acquired skills, (3) take his break on time, and (4) perform the job safely. However, his primary goal at the moment he depressed the circuit breaker button was to (3) take his break on time. Though his choice almost killed someone, you must look carefully and dispassionately at the motivations surrounding his choice, without regard to the consequences, remembering that people don't come to work to fail.

The *behavior choice assessment* provides a logical means of "adjudicating" a person's choices made at the time of the **event**.[58] The assessment, illustrated in Figure 7.6, is derived largely from Dr. David Marx's books, *Whack-a-Mole* and *Dave's Subs*,[59] which David refers to as the "Simplified Just Culture Model."

Ultimately, the assessment helps you focus your limited resources where they will do the most good. Simply blaming a hapless electrician for an **event** does very little to improve not only his future performance but also the organization's resilience and reliability.

The Simplified Just Culture Model helps you "understand" the choices of those whose behaviors do not align with organizational values or procedural rules. It can be used to appraise anyone's choices, whether executive or front-line worker, regarding any unwanted occurrence. With it, you can determine which of three behaviors was most likely in play. This gives you the ability to address the incident and the people involved in a constructive way rather than simply reacting to the outcome. In particular, you want to distinguish between the following behavior choices:

- *Human error*—it's normal for people to err—it happens to the best of us. Slips, fumbles, lapses, trips, mistakes, etc. are par for any human being. Actions performed in error become opportunities to learn and to improve our systems—presence of error traps and defenses that failed to protect assets. Most people are horrified that their mistakes triggered harm— mainly, they need consolation and encouragement. Any system that is one failure away from harm, be it human error or equipment failure, is vulnerable.
- *At-risk choice*—people sometimes get overconfident and drift away from the rules (like driving a few MPH over the speed limit, for example); they make choices that could place themselves and others in harm's way. Often, when people "transgress" rules, policies, or procedures, they think they are "in control, and "it can't happen to me." Or, they may feel it's "justifiable." Additionally, it may be the way "we've always done it," but nothing bad happened before now. People in general think in terms of likelihood and not consequences—they don't "see the risk." Coaching and education are the best responses to the individual, reminding them of the risks that may have been forgotten or mistakenly justified. There are also opportunities for organizational learning.
- *Reckless behavior*—in very rare occasions, people choose intentionally to place themselves or others in harm's way. They know the potential consequences. They know the safer course of action but deliberately choose not to adhere to it. They elevate their own self-interests above others. An individual who makes reckless choices, needs to be subject to disciplinary sanctions.

Human error and "one-off" at-risk choices are starting points, not end points for **event** analysis. In both cases, there was no intent to either cause harm or put anyone in harm's ways. And, there are system weaknesses making the system vulnerable to everyday mis-steps or seemingly "logical" choices made for the sake of efficiency. In most cases, accountability occurs when people simply tell their story—what could or should have been done is usually obvious.

Important! The *behavior choice assessment* can be used effectively, avoiding a blaming-and-shaming reaction to human error, while preserving accountability. Users, especially managers, must understand and apply systems thinking to explain an individual's performance.

This tool helps analysts and managers avoid hindsight bias and related emotional reactions. Hindsight bias is the mistaken belief developed by an **event** analyst after the fact that an outcome "could have" been readily predicted by the individual(s) involved before the **event** occurred. "They should have seen this coming." This bias, formed by an analyst who has information after the fact that the person did not have or appreciate at the time, inevitably works against treating workers fairly. However, by systematically and objectively understanding the individual's intentions at the time, hindsight bias can be minimized, if not, eliminated. The behavior choice assessment helps managers avoid jumping to conclusions based on emotional reactions by providing a systematic means of:

- identifying actions and the intent of those actions;
- moderating managers' emotional response, emphasizing the facts;
- providing opportunity for individuals to tell their stories;
- sustaining the flow of information between the workforce and management.

As you can tell, conducting a *behavior choice assessment* promotes both accountability and learning at the same time, which is the intent of a just culture (discussed later in Chapter 8). In most cases, people stand ready to give an account of that for which they are responsible—they know accountability is necessary to keep people honest. Most people come to work to do a good job—not to fail— but want to be treated fairly and with respect. This is why David Marx refers to the algorithm as a "just culture" model. If a performer is "innocent" of reckless motives and shoddy work, the individual is considered free of blame, and the organization's responsibility and role in the **event** become self-evident.

In reference to Figure 7.6, after pinpointing the actual behavior that triggered the **event**—the harm of that action (or threatened harm in the case of a near hit), the analyst considers, in order, recklessness, at-risk choice, and human error. The first two questions address recklessness. Questions 3 and 4 resolve at-risk choices and human error.

Question 1 assesses the person's purpose. In very rare cases, the person who triggered an **event**, intentionally instigated the **event** for the purpose of causing substantial harm—a serious offense to the organization and, perhaps, society. In such cases, the person deserves sanctioning.[60] In most cases the answer to this question is no. Question 2 looks carefully at the "justification" for the person's choice.

Action (that threatened harm): _____

Harm (threatened by action): _____

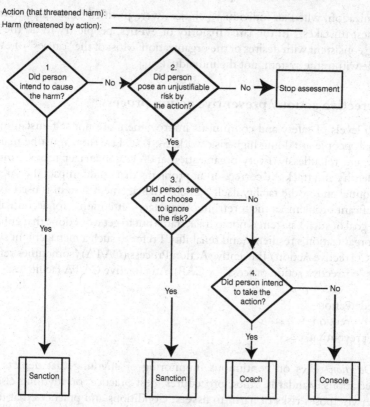

Figure 7.6 The behavior choice assessment based on the "Simplified Just Culture Model." This assessment is useful for selecting a just and dignified response to an individual's choices. (Source: Personal communication with David Marx. Used with permission. Copyright (2017) Outcome Engenuity, LLC.)

Again, on rare occasions, it may be necessary to take serious risks, such as attempting to save the lives of others. There may be a reasonable and logical explanation for the choice at the time for which the person was "justified" in making. If so, there is no longer a reason to continue the behavior choice assessment. However, if the person mistakenly thought they were justified or misjudged the risk, the answer to Question 2 is yes. Now, you ask Question 3. Did the person see and choose to ignore the risk? If yes, again, the proper response is sanction. Most of the time, the answer to this question is no, which takes you to Question 4. A yes response to this question suggests a lack of knowledge, skill, or understanding of the situation. Coaching or training is a reasonable response to this situation. If the answer to Question 4 is no, then consolation is the best path for you as the manager to take. Most people feel terrible for their mistakes, especially those that triggered harm.

All in all, the *behavior choice assessment* tool helps the user look unemotionally at the facts and to shift the focus from the sharp end to the blunt end of the organization, when an individual's choices were justified or simply in error—"honest mistakes." In the large majority of **events**, people try to do the right thing consistent with desires of the organization. Most of the "causes" of **events** reside within the system, not the individual.

Corrective action / preventive action process

High levels of safety and continuous improvement are not self-sustaining. As long as people are doing high-risk work, SYSTEMS LEARNING must be managed daily and relentlessly. Every organization and every manager needs a method to identify and track the correction of problems that could impact the safety of personnel and/or the facility itself.[61] Such a management system is used to keep significant problems at the forefront of managers' attention—for accountability. The goal of such a system is not to track status but to get work done that enhances the organization's resilience and reliability. I refer to such a management system as a Corrective Action / Preventive Action Process (CA/PA) (sometimes referred to as "corrective action process" or CAP). An effective CA/PA facilitates:

- detection
- correction
- prevention.

Detection relies on continuous monitoring of anything that departs from expectations, standards, rules, procedures, best practice, or anything else that otherwise poses risks of harm to **assets**. Conditions and practices that do not conform to quality and regulatory requirements are recorded for and tracked to resolution. Reports of problems—including suggestions for improvement—are documented and reviewed promptly for safety, reliability, quality, productivity, efficiency, and regulatory requirements.

> *Warning!* Once problems are recognized, the search for causes must always precede the determination of corrective/preventive actions. You cannot know where your systems need tuning until you know the how's and why's.[62] Otherwise, implementing such actions before causes are known could lead to unintended outcomes and unneeded expenses.

Correction depends on thorough analysis of problems—including the review of *local* and *organizational factors* contributing to the issues (misalignments). Problems are corrected consistent with their importance and significance to safety or mission. Sound corrective actions are more organizational and systemic in nature. Relative to **Hu**, do not admonish front-line workers to simply "pay attention." Also, be wary of training and procedures as the default corrective actions for **events** related to **Hu**. Both are expensive. Training should be used only when there are knowledge gaps or skill deficiencies important for job performance. Procedure revisions may be necessary, but understand that more rules do not necessarily equate to more safety. Remember, a procedure is a control—a defense that guides the user's behavior choices. Just as a map is not the terrain, a procedure is not the work.[63]

Senior managers, who monitor significant problems to ensure timely resolution of vulnerabilities, emphasize *prevention*. This involves preparation associated with anticipation for expected activities as well as follow-through to repair minor issues before they reveal themselves as **events**.

Generally, an effective CAPA management system institutionalizes detection, correction, and prevention into the following functions:

1   *Identification.* Any person with access to the company intranet can report a problem that documents any issue—including work planning, after-action reviews, operating experience, self-assessments, benchmarking, adverse trends, learning team results, observation follow-up, etc. This function documents a problem to ensure it is tracked to resolution.

2   *Screening.* This is a structured review of problems that require immediate response (safety, productivity, regulatory, etc.)—including immediate and interim corrective actions until fundamental issues can be determined and corrected.

3   *Assignment.* Someone is given responsibility and sponsorship for resolving the problem—based on knowledge, technical expertise, experience, and budget jurisdiction.

4   *Prioritization.* Managers decide on the urgency of response. *Caution! If at all practicable, don't live with problems.*

5   *Evaluation.* An analysis of system weaknesses and defective defenses contributing to the problem, including a determination of extent of conditions, is conducted.

6   *Corrective actions.* Managers identify needed improvements and interventions related to organizational processes, management systems, and other systemic issues—including assigning responsibilities, development of corrective action plans, resource allocation, and specification of milestones and deadlines.

7   *Feedback.* Information is provided to the person(s) making the original report as to management's response to the problem and progress on its resolution.

8   *Trending and monitoring.* The coding of issues and long-term tracking of metrics helps identify repeat **events**, recurring problems, and subtle organizational issues not previously detected. Periodic summary reports are submitted to line managers for collegial review.

9   *Effectiveness.* The management team reviews the success (resolution) of corrective actions for significant problems.

10 *Closure.* This is an explicit recognition that a problem is solved and corrected—a celebration might be appropriate!

A CA/PA process cannot be considered optional. Such an important management system must be highly valued by the organization's executives and senior managers. It's usually institutionalized with a written policy and related process procedures that include regular review of CA/PA reports. Without visible support and accountability, line managers will tend to perform CA/PA-related functions in a perfunctory manner or ignore them altogether. I've seen this happen. Line managers are responsible for not only production but also safety and resilience. CA/PA keeps them honest about the true state of their management systems. Executives and senior managers signal their organizational priorities and desires by what they pay attention to and by what questions they ask. As emphasis on SYSTEMS LEARNING diminishes, correction of system weaknesses declines. When these weaknesses persist—more vulnerabilities increase—accumulation occurs, which accelerates the incubation of your next **event**.[64]

The **event**, "First impressions are lasting impressions," described in Chapter 4, demonstrates the reality of how **events** emerge from a poorly maintained organizational system. For sure, Kelly used poor judgment when he took the 50–50 chance of being right and pushed a button. Yet the system set him up for just such a choice.

- An inflexible work culture pushed workers to maintain an inflexible schedule for taking routine breaks. Hence, Kelly's fear of his supervisor's "take-it-or-lose-it" mandate toward break time fed his desire to act quickly.
- A strong emphasis on schedule adherence left a relatively new employee feeling reluctant to stop and ask for help about his conundrum. A more safety-conscious organizational climate would have addressed the permission to stop when unsure during Kelly's orientation and training. When safety is in question, people need explicit encouragement to know that delays are acceptable—especially in the context of making sure they're doing the right thing.
- Financial pressures that led to reduced staffing decisions, combined with schedule pressures, led to reducing the time available for training, such that Kelly had never been given the opportunity to actually perform the task before being allowed to work unsupervised.
- A management team that tolerated such problems as worn-out labels on critical controls set a trap in which Kelly had only a 50 percent chance of making the correct decision—open or close. It amounts to a willingness to let one worker gamble with another's well-being.

These system deficiencies created the perfect set-up for a new employee—intent on doing a good job—to make a serious mistake. The system collided with the worker's frustration and fear of being viewed as unqualified, and it resulted in an extremely serious near hit—someone could have died. A determined, relentless management posture toward SYSTEMS LEARNING, if it had existed, would have identified and corrected conditions like these, possibly preventing similar **events**.

Top-performing managers actually welcome a CA/PA process as an important management tool to ensure continuous system improvement. Continuous improvement in H&OP—SYSTEMS LEARNING—is mandatory to sustain safety and resilience—it's not optional. If learning diminishes, you will eventually learn through **events**. Without CA/PA and the accountability that goes with it, harm is inevitable.

Learn from success as well as failure

A typical approach to fixing problems in a system is to evaluate where things have gone wrong in the past, and determine what changes are needed to keep it from happening again. While this is not an unreasonable or inappropriate strategy for improvement, it is not a complete solution, either. In his book, *Safety-I and Safety-II*, Dr. Hollnagel points to a more holistic perspective with his Safety-II guidelines for improving the resilience of complex organizational systems.[65]

"When things go wrong, we often seek to find and fix the 'broken component', or to add another constraint. When things go right, we pay no further attention."

—Erik Hollnagel
Author: *Safety-I and Safety-II* (2014)

"Safety-I thinking" focuses on fixing what went wrong and containing the effects of what goes wrong. When an **event** occurs, it only makes sense to see what happened. We assume that if we can replace broken components, eliminate the actions and/or conditions that triggered the **event**, we tend to think it won't happen again. That is true to an extent. The problem with Safety-I thinking is that an **event** must occur in order to discover the problem and its corrective actions.[66] Something must go wrong before we can figure out how to keep it from happening again. And since things do go right most of the time, there are few **events** to be studied. If we leave our system integrity at the mercy of **events**, the probability for improvement—and the prevention of future **events**—is lost. SYSTEMS LEARNING is both proactive and reactive—but it is most effective when it is proactive.

Here's where "Safety-II thinking" steps in. Hollnagel points out that our ability to create better systems is greatly enhanced if we examine what goes right—in addition to studying what goes wrong. It emphasizes optimizing the number of successful outcomes, making sure things go right. He points out that safety is something that happens, not something that does not happen.[67] If we understand why things go right—why most products and services meet customer requirements every day—we foster thinking about how to make sure we keep on doing the right things right—CRITICAL STEPS and RIAs.

The benefit of analyzing what goes right also lies in the quantity of data available. Even at a standard human reliability rate of 99 to 99.9 percent (which you'll recall isn't good enough for CRITICAL STEPS), only one failure will occur in a thousand occasions. That means System I evaluation has one **event's** worth of data to review, while System II tends to look at the 999 occurrences that go right. Remember that *work-as-done* rarely matches *work-as-imagined* exactly. Therefore, learn why things go right—people regularly make adjustments to accomplish their goals. Adding System II thinking to your field observations and self-assessments takes advantage of one of human nature's positive factors: people much prefer to talk about what they've done right than what they've done wrong.

When managers commit to examining their organizations this way, they can create a system that learns from its mistakes and incorporates ongoing feedback—integrating change into all work stages.

Things you can do tomorrow

1   Ask yourself if people feel free to talk to managers and executives about problems—do you create conversations (dialogue) with front-line workers? Are people rewarded if they spot potential problems?

2   Take a walk around the production areas and talk with people. Ask them what they are trying to accomplish and what obstacles they regularly encounter. Ask them if procedures can be followed as written—follow this question with queries about what is normally done. Ask them if training is sufficient for what they are asked to do.

3   Ask subordinate managers and supervisors to spend time on the shop floor or in the work spaces at least weekly to observe work as it's really done, and ask for their conclusions during the next management meeting.

4   Observe a frequently-occurring work activity concurrently with a subordinate manager or supervisor, observing basic work practices and work conditions. Afterward, compare what the other person heard and saw with what you heard and saw to identify important differences in conditions observed and adherence to expectations, standards, and feedback methods.

5   Ensure experienced, proficient front-line workers provide technical input in important operational decisions.

6   Direct the responsible manager for a unit or organization involved in a serious **event** to personally present the results of the **event** analysis to the management team. Does the manager take a systems approach or a person approach?

7   For your next **event** analysis, assume it is an organizational failure instead of a failure of the individual who triggered it. Why didn't the organization know about the unsafe conditions before the **event**?

8   During **event** analysis, encourage the involved personnel to share their stories—about how things evolved over time as they perceived them. Understand the work situation from their point of view.

9   Incentivize certain leading indicators, such as the number of CRITICAL STEP MAPPINGS completed, field observations conducted, procedures and trainings verified during "dry-runs" before starting high-risk work, pre-

job briefings conducted, etc., which measure the presence of safety—if incentivized, they foster RISK-BASED THINKING.

10   Analyze your **event** and near-hit data from the last two to three years to identify (1) the most frequent behavior choices that led to those outcomes, and (2) the most common failures in defenses and their system weaknesses.

Notes

1   A frequent refrain of Dr. William Corcoran of Nuclear Safety Review Concepts in his *Firebird Forum* newsletter.

2   *Economics of Defense Policy: Adm. H. G. Rickover. Part 2, Selected Congressional Testimony and Speeches by Adm. H. G. Rickover, 1953–81*. Hearing Before the Joint Economic Committee Congress of The United States Ninety-Seventh Congress, 2nd Session (1982) (p.23).

3   Federal Highway Administration, U.S. Department of Transportation. Highway Statistics Series. Retrieved 9 April 2017 from https://www.fhwa.dot.gov/policyinformation/statistics.cfm.

4   Groeneweg, J. (2002). *Controlling the Controllable: Preventing Business Upsets* (5th edn). Global Safety Group (pp.246–247).

5   Spear, S. (2009). *The High-Velocity Edge*. New York: McGraw-Hill (pp.90–93).

6   None of the events involved actual damage to the reactor core. The risk of core damage was assessed based on the application of probabilistic risk assessment (PRA) methods.

7   Idaho National Engineering and Environmental Laboratory (March 2002). "Review of Findings for Human Contribution to Risk in Operating Events," (NUREG/CR-6753). Washington, DC: U.S. Nuclear Regulatory Commission (p.xi).

8   Snook, S. (2000). *Friendly Fire: The Accidental Shootdown of U.S. Blackhawks Over Northern Iraq*. Princeton: Princeton University Press (p.24).

9   Leveson, N. (2011). *Engineering a Safer World: Systems Thinking Applied to Safety*. Cambridge, MA: MIT (p.274).

10 Vaughn, D. (1996). *The Challenger Launch Decision: Risky Technology, Culture, and Deviance at NASA*. Chicago, IL: University of Chicago (pp.62–68).

11 Viner, D. (2015). *Operational Risk Control: Predicting and Preventing the Unwanted*. Farnham: Gower (p.169).

12 The concept of accumulation was described by Barry Turner and Nick Pidgeon as an *incubation period* during which attitudes, beliefs, practices, and defenses eroded unnoticeably by an organization. See their book, *Man-Made Disasters* (1997). (2nd edn) Oxford: Butterworth/Heinemann (pp.68–84).

13 Federal Railroad Administration (FRA) (2008). Risk Reduction Program: A New Approach for Managing Railroad Safety. Retrieved 27 May 2017 from http://www.fra.dot.gov/downloads/safety/ANewApproachforManagingRRSafety. pdf.

14 Hudson, P., Verschuur, W., Lawton, R., Parker, D. and Reason, J. (1997). *Bending the Rules II: The Violation Manual*. Leiden: Rijks Univeriteit Leiden.

15 Hudson, P. et al. (2000). "Bending the Rules: Managing Violation in the Workplace," *Exploration and Production Newsletter*, EP2000-7001 (pp. 42–44). The Hague: Shell International. Though there are many references related to compliance/non-compliance with procedures and expectations, this illustration of the factors enabling/inhibiting non-compliance and drift was inspired by Rob Fisher's discussion on *deviation potential*. See https://www.wecc.biz/Administrative/2013%20HPWG%20Session%201%20Elements%20of%20an%20Effective%20HPI%20Program%20(Rob%20Fisher).pdf. This relationship was also influenced by the paper by Hudson, P. et al.

16 Pittet, D. (April 2001). Improving Adherence to Hand Hygiene Practice: A Multidisciplinary Approach. Centers for Disease Control and Prevention (Reviewed 10 May 2011). Retrieved 27 May 2017 from http://wwwnc.cdc.gov/eid/article/7/2/70-0234_article.

17 Vaughn, D. (1996). *The Challenger Launch Decision: Risky Technology, Culture, and Deviance at NASA*. Chicago, IL: University of Chicago (p.407).

18 Hudson, P., Verschuur, W., Lawton, R., Parker, D. and Reason, J. (1997). *Bending the Rules II: The Violation Manual*. Leiden: Rijks Univeriteit Leiden.

19 LA Times article about animal attacks. Retrieved 26 May 2017 from http://articles.latimes.com/2002/may/31/world/fg-crocs31.

20 Hofman, D. and Frese, M. (2011). *Errors in Organizations*. New York: Routledge (p.217).

21 Hofman, D. and Frese, M. (2011). *Errors in Organizations*. New York: Routledge (p.233).

22 *Deference to expertise* is one of five cardinal tenets of High-Reliability Organizations (HRO). Deference to expertise encourages managers to let the people with the most expertise, not just with the highest rank, make operational decisions—letting operational decisions "migrate" to those with the technical expertise to make them.

23 Reason, J. (2008). *The Human Contribution: Unsafe Acts, Accidents and Heroic Recoveries*. Farnham: Ashgate (p.241).

24 I credit Dr. Bill Corcoran's wisdom on this important communication function.

25 Reason, J. (1997). *Managing the Risks of Organizational Accidents*. Aldershot: Ashgate (p.198).

26 Ibid. (p.195).

27 Dekker, S. (2007). *Just Culture: Balancing Safety and Accountability*, Farnham: Ashgate (p.43); Marx, D. (2009). *Whack-a-Mole*. Plano, TX: By Your Side Studios (p.116).

28 See U.S. NRC guidance on establishing and maintaining a safety-conscious work environment at: http://www.nrc.gov/about-nrc/regulatory/allegations/scwe-mainpage.html.

29 I want to point you to two important resources on the topic of just culture. David Marx's book, *Whack-a-Mole*, is outstanding and practical. Dr. Dekker's book, *Just Culture*, takes a more academic and legalistic view of accountability. Both references provide insights important to creating and sustaining high reliability and resilience by treating people with respect and dignity.

30 Taiichi Ohno, the Toyota executive most responsible for developing the Toyota Production System, was known to draw a chalk circle on the floor around a manager and would make them stand inside the circle until they had noted all of the problems in a particular production area. This lean technique is known as the "Ohno circle."

31 Kletz, T. (2001). *An Engineer's View of Human Error*. New York: Taylor and Francis (p.104).

32 Conklin, T. (2012). *Pre-Accident Investigations: An Introduction to Organizational Safety*. Farnham: Ashgate (p.43–44).

33 Reason, J. (1998). *Managing the Risks of Organizational Accidents*. Aldershot: Ashgate (p.21).

34 Hollnagel, E. (2014). *Safety-I and Safety-II: The Past and Future of Safety Management*. Farnham: Ashgate (p.161).

35 Hollnagel, E. (2014). *Safety-I and Safety-II: The Past and Future of Safety Management*. Farnham: Ashgate (p.172).

36 John Wreathall, author of several editions of books on *Resilience Engineering* and a specialist on organizational metrics, says this about interpreting trends: "I cannot emphasize enough the need *not* to pursue every tick in the data—it's the fastest way to burn out any program however well intentioned."

37 Hollnagel, E. (2014). *Safety-I and Safety-II: The Past and Future of Safety Management*. Farnham: Ashgate (p.12).

38 Reason, J. (2008). *The Human Contribution*. Farnham: Ashgate (p.139).

39 Reason, J. (2013). *A Life in Error: From Little Slips to Big Disasters*. Farnham: Ashgate (pp.78–80).

40 Spitzer, D. (1997). *Transforming Performance Measurement: Rethinking the Way We Measure and Drive Organizational Success*. New York: AMACOM (p.95).

41 Wreathall, J. (2009). Measuring Resilience. In Nemeth, C., Hollnagel, E. and Dekker, S. (Eds.). In *Resilience Engineering Perspectives, Volume 2: Preparation and Restoration*. Farnham: Ashgate (pp.112–114).

42 Viner, D. (2015). *Operational Risk Control: Predicting and Preventing the Unwanted*. Farnham: Gower (p.118).

43 Hollnagel, E. (2004). *Barriers and Accident Prevention*. Aldershot: Ashgate (pp.98–99).

44 A "line of position" is a bearing line drawn between a fix point of reference on a nautical or aeronautical chart and a moving ship or aircraft. Own ship's position is somewhere along that line. The distance from the craft to the fixed point is unknown until a second line of position (bearing) is taken on another, separate fixed point on the chart. The intersection of the two lines of position (a point) fixes the location of the ship or aircraft on the chart. A third line of position (dot or small triangle) increases the reliability of the craft's location on the chart—a geographic representation of the real world.

45 Reason, J. (2008). *The Human Contribution: Unsafe Acts, Accidents and Heroic Recoveries*. Aldershot: Ashgate (pp.138–139).

46 Dekker, S. (2011). *Drift Into Failure: From Hunting Broken Components to Understanding Complex Systems*. Farnham: Ashgate (p.181).

47 Dennis Murphy, a former U.S. nuclear submarine commanding officer, thinks about event analysis this way: Think about problems and events in terms of basic geometry: (1) a *point solution* (one dimension) is simple, narrow and targets an immediate fix to a symptom (i.e. a Band-Aid to stop the bleeding); (2) a *plane solution* (two dimensions) considers multiple causal factors but falls short of addressing underlying systemic issues that, left unchecked, can lead to recurrence of the problem; and finally, (3) a *spherical solution* (three dimensions) that considers the entirety of interdependent systems (and processes), which conspired together to deliver an undesired outcome.

48 The idiom, "red herring," is something that diverts attention from the real problem or matter at hand—a misleading clue. Reliance on "human error" as a cause diverts attention as well as resources from the fundamental systemic weaknesses higher up (or deeper) in the organization.

49 Reason, J. (1997). *Managing the Risks of Organizational Accidents*. Aldershot: Ashgate (p.17).

50 Maurino, D., Reason, J., Johnston, N. and Lee, R.B. (1995). *Beyond Aviation Human Factors*. Aldershot: Ashgate (p.66).

51 Dekker, S. (2014). *The Field Guide to Understanding 'Human Error'* (3rd edn). Farnham: Ashgate (p.6).

52 Dekker, S. (2006). *The Field Guide to Understanding 'Human Error'* (3rd edn). Farnham: Ashgate (p.8).

53 Dekker, S. (2006). *The Field Guide to Understanding 'Human Error'* (3rd edn). Farnham: Ashgate (pp.46–48).

54 Johnson, W. (1980). *MORT Safety Assurance Systems*. New York: Dekker (pp.74–85).

55 Chevalier, R. (2007). *A Manager's Guide to Improving Workplace Performance*. New York: American Management Association (pp.101–112).

56 Agnew, J. and Daniels, A. (2010). *Safe by Accident? Take the Luck out of Safety*. Atlanta, GA: Performance Management Publications (p.129).

57 Reason, J. (1997). *Managing the Risks of Organizational Accidents*. Aldershot: Ashgate (p.211).

58 I use the term "adjudicate" here to emphasize the decision-making regarding one's motives. It's a judgment regarding the motives behind a specific choice made in the context of an **event**.

59 Marx, D. (2007). *Whack-a-Mole: The Price We Pay for Expecting Perfection*. Plano, TX: By Your Side Studios. And (2015). *Dave's Subs: A Novel Story of Workplace Accountability*. Plano, TX: By Your Side Studios.

60 Sanction involves a range of organizational responses to the person designed to inhibit the unacceptable behavior choice in the future as well as serve as threat to others who may be tempted to make similar choices.

61 Subalusky, W. (2006). *The Observant Eye: Using It to Understand and Improve Performance* (p.11). (Self-published at booksurge.com.) A *problem* is anything that needs correction—a gap between what is and what should be, including any activity not performed according to expectation or best practice and any condition that should not exist, which poses a greater risk.

62 Marx, D. (2015). *Dave's Subs: A Novel Story of Workplace Accountability*. Plano, TX: By Your Side Studios (p.185–187).

63 Dekker, S. (2015). *Safety Differently: Human Factors for a New Era* (2nd edn). Boca Raton, FL: Taylor and Francis (p.103).

64 Ramanujam, R. and Goodman, P. (2011). The Link Between Organizational Errors and Adverse Consequences: The Role of Error-Correcting and Error-Amplifying Feedback Processes. In Hofmann, D. and Frese, M. (Eds.). *Errors in Organizations*. New York: Taylor and Francis (p.262).

65 This discussion draws from Hollnagel, E. (2014). *Safety-I and Safety-II: The Past and Future of Safety Management*. Farnham: Ashgate (p.183).

66 Viner, D. (2015). *Occupational Risk Control: Predicting and Preventing the Unwanted*. Farnham: Gower (p.117).

67 Hollnagel, E. (2014). *Safety-I and Safety-II*. Farnham: Ashgate (p.136).

8   Managing human performance

> The behavior of people in business is not another issue to be considered—it is at the center of every business decision.
>
> Aubrey Daniels[1]

> In the final analysis, we must depend on human beings. No machine, including a computer, can be more perfect than the human beings who designed it, use it, or rely on it.
>
> Hyman Rickover[2]

Success is realized when your organization both accomplishes its business needs (what you want) and avoids unwanted outcomes (what you don't want). Success is achieved through behavior. Nothing good or bad happens until someone does something to initiate a transfer of energy, movement of mass, or a transmission of information. Work is behavior. This is what Dr. Daniels meant by his quote above.

As a follow-up to the previous chapter on SYSTEMS LEARNING, the purpose of this chapter is to describe what a manager can do to get and sustain the desired behaviors and practices from individuals—people following procedures, adhering to expectations, using **Hu** tools, and adapting to changes in high-risk situations to protect **assets** from harm. **Hu** is best described as the accomplishment of results through an individual's choice of behaviors. Before embarking on an initiative to integrate H&OP into your operations, it's important to target the new behaviors that you want and the ones you don't want. You should recall that behavior is the linchpin of both the *alignment* and the *systems thinking models*, which should guide your thinking about **Hu**.

In the nuclear industry, managers generally attempted to "manage" performance through training, procedures, and supervision—what is widely known as the three-legged "nuclear model." Though this approach does improve an organization's resilience, it is an overly simplistic view of **Hu** that doesn't consider the hundreds of other factors influencing choices in the workplace. Hence, the focus of this chapter is on an engineering approach (systematic and repeatable) that promotes desired performance from front-line personnel by (1) specifying expectations, (2) identifying and diagnosing performance gaps (or opportunities), (3) identifying effective interventions that will sustain desired behaviors consistent with the organization's business needs, and (4) aligning the organization to establish the conditions that enable desired behaviors and inhibit unwanted practices.

Let me be clear. **Hu** is management's responsibility. You cannot delegate this responsibility to a staff member who "specializes" in human performance. The actions that people do in the workplace are in response to your specific direction and the system you create. A manager's primary value is the ability to influence and sustain behaviors of people at the sharp end—what people do. Therefore, whether an executive, manager, or supervisor, you must learn to "engineer" the organization—the system—to promote and sustain the behaviors and results you want and to eliminate the behaviors and practices you don't want. This chapter, piggy-backing on learning methods described in Chapter 7, will provide insights in how to manage **Hu**.

Dr. James Reason suggested that it is easier to manage the conditions people work in than to manage the human condition. But, Reason goes on to say that "free will is an illusion because our range of actions is always limited by the local circumstances."[3] As a manager and the owner of the system, you have a great deal of influence over the behavior choices people make in the workplace through workplace *local factors*, and you have leverage over most of them. Collectively, they influence the worker's perception of **pathways** and CRITICAL STEPS. Recall that these factors are a blend of conditions created by the system itself via the various organizational functions, management systems, and corporate values (avenues of influence). Together, they drive individual worker behavior.

When RISK-BASED THINKING and H&OP (control, learn, and adapt) have been integrated into your organization, you should see many of the behaviors listed in Table 8.1 occurring regularly throughout the organization.

Basic management cycle

I hope this paragraph does not offend you, but I believe it's an important reality that must be addressed—whether it applies to you or the managers and supervisors who report to you. Here's a sad reality that I have discovered. Just because the term manager appears in someone's official title, that person may not necessarily know how to manage. In most cases, the person was promoted to a supervisory or managerial position because of their technical experience, usually without the requisite knowledge or ability on how to manage or supervise people, much less manage **Hu**.

I don't want to diminish the importance of leadership in human performance. Leadership skills are key to motivating an organization's personnel in achieving both its mission and its vision. But I have discovered that line managers frequently have little understanding of how to instill a new or change workgroup behavior. The word "management" suggests a systematic approach to controlling activities in order to achieve a goal. To manage something involves:

Table 8.1 Evidence of integration. Certain practices will generally exist throughout an organization when RISK-BASED THINKING and H&OP have been successfully integrated into operations (caution: this is not a complete list)

Executives / Managers / Supervisors	Front-line workers

• Ongoing integration of RISK-BASED THINKING into management systems and operational processes.
• Using the H&OP vocabulary during meetings.
• Identifying CRITICAL STEPS in operations using CRITICAL STEP MAPPING.
• Applying a variety of systems learning methods.
• Encouraging accountability for SYSTEMS LEARNING, using a CA/PA process.
• Conducting observations of work in progress.
• Giving feedback:.
  ° reinforcing desired behavior
  ° coaching deficient behavior
  ° correcting unacceptable behavior
  ° consoling unintended behavior.
• Soliciting and accepting feedback with an attitude of humility from front-line workers.
• Developing and following through on expectations .
• Conducting local factor analysis for new expectations.
• Conducting local rationality **event** analysis.
• Acknowledging positive values and principles.
• Preserving the dignity and respect of people.
• Promoting a systematic approach to technical training.
	• Adopting RISK-BASED THINKING to adapt conservatively to novel situations that require timely responses.
	• Identifying CRITICAL STEPS and related Risk-Important Actions for assigned tasks.
	• Identifying assets and related hazards for assigned tasks.
	• Applying selected **Hu** tools, such as:.
	  ° pre-job briefing
	  ° self-checking
	  ° stop when unsure
	  ° conservative decision-making.
	• Recognizing and pausing to think before:.
	  ° transfers of energy.
	  ° movements of mass.
	  ° transmissions of information.
	• Giving and receiving verbal feedback (willing and able to challenge bosses and co-workers); not taking offense when given feedback by co-workers or less experienced personnel.
	• Reporting significant differences in *work-as-done* and *work-as-imagined* (planned) using most effective communication methods including verbal, written, and digital.

1   Knowing where you want to be and where you are.
2   Developing a plan to close the gap.
3   Implementing the plan.
4   Monitoring progress.
5   Adjusting as needed to further close the gap.

Because of its effectiveness and simplicity, W. Edward Deming's "Plan–Do–Check–Act" management cycle[4] offers a sound mental framework for managing H&OP. With one minor revision to the Deming management model, the essential elements of managing include the following steps:[5]

1 *Plan*—know where you are, where you want to be, and the steps to get there. Form clear expectations. Define the gap between expectation and actual performance, including the means to close the gap.
2 *Do*—execute the plan. Communicate, train, and implement the plan. Align the organization—the system—to enable the desired performance and inhibit unwanted practices.
3 *Check*—observe progress toward the goal. Monitor the effectiveness of implementation of expectations and accomplishment of outcomes—look for performance gaps.
4 *Adjust*—collect relevant feedback and adapt the plan or expectation. Alter the plan or expectations and means to more closely achieve the desired performance.

A performance gap is simply a noticeable difference between the results you want and the results you are getting—between actual behavior and expected behavior—between current practice and preferred practice. However, before proceeding with the performance management process, you must confirm that the actual behavior (*work-as-done*) is indeed undesirable. It may actually be what you want, given the circumstances encountered by the performer. The choices people made at the time may have involved an adjustment to an unanticipated set of conditions that posed a risk to one or more **assets**. And, in order to avoid harm, the person took appropriate action to protect **assets**. In this case, praise the individual for his/her adaptation. Then, try to understand how your system created such a risky situation.

A performance gap is a performance problem, which needs a solution—closure. This could be the result of drift, which, as you'll recall, is a long-term phenomenon. *Work-as-done* may indeed be a long-term regression in performance from expectation (*work-as-imagined*). Alternatively, the performance gap may result from a change in technology, a new installation or equipment modification, lessons learned from a previous **event**, or other changes in working conditions, procedures, processes, or tools. Either way, there is a performance problem that needs your attention.

Caution! New employees of Nordstrom Department Stores were handed a card to keep in their possession, which included Nordstrom's one rule: "Use good judgment in all situations."[6] Expectations are always formed with assumptions in the mind of the originator and can add complexity to work. So, when novel situations or confusion arises, or when assumptions cannot be met, the front-line worker must be free to "use good judgment."

If you didn't already notice it, basic management is resilience in action. The plan–do–check–adjust management cycle closely aligns with the cornerstone practices of RISK-BASED THINKING: anticipate (plan), monitor (check), respond

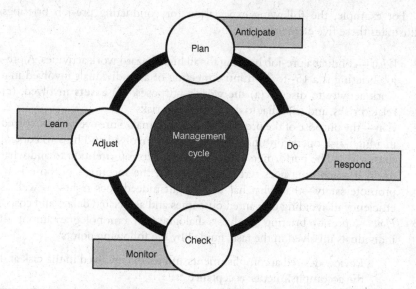

Figure 8.1 The management cycle. Effective managers follow an iterative cycle toward accomplishing their goals, whether for accomplishing production goals or safety goals. As a form of management, the four habits of thought of RISK-BASED THINKING are superimposed behind the management cycle

(do), and learn (adjust) (see Figure 8.1). Effective managers more readily adopt and more easily practice RISK-BASED THINKING habits of thought in concert with their management responsibilities.

Plan—form expectations

An expectation is the anticipation of achieving a desired accomplishment—a strong belief that someone will or should do something. More simply stated, it's results a manager wants from his people—*work-as-imagined*. An expectation is a specific statement of an accomplishment or work output, along with its criteria for success, and, if necessary, the relevant behavior(s) required to safely achieve the desired work output. You should recall that "Expectations and feedback" is cell 1 of the *local factors* table. Expectations and feedback is #1 for a reason—it represents the greatest point of leverage in your management system. An expectation has the following five elements:

- *What*? The accomplishment/result/work outputs/tolerances of acceptability.
- *Why*? Business case / reason for it / protection of **assets.**
- *How*? The behavior(s) required to achieve the desired work output (to minimize risk).
- *When*? During what tasks / situations / conditions / periods?
- *Where*? Location / workplace.

For example, the following expectation for conducting pre-job briefings illustrates these five elements:

1 *What*—conduct a pre-job briefing for all high-hazard work activities. A pre-job briefing is a 15- to 30-minute meeting of all individuals involved in a work activity to discuss (a) the work's purposes, (b) **assets** involved, (c) related risks, and (d) what to do about those risks.

2 *Why*—the purpose of the pre-job briefing is to make sure everyone involved in a high-risk operation understand the risks involved and how to control those risks. The performers understand what to pay attention to and what to do if something goes wrong. Ultimately, the meeting's purpose is to promote safety—that what has to go right indeed goes right—as well as efficiency in avoiding unwanted outcomes and associated delays and costs.

3 *How*—a pre-job briefing involves a dialogue (not a monologue) among all individuals involved in the task, guided by the following points:

   i Review desired accomplishments and **assets** involved in the task and the accomplishments' acceptance criteria.

   ii Understand the intrinsic **hazards** involved in the work and previous experiences with the task, including previous **events**.

   iii Summarize all CRITICAL STEPS and related Risk-Important Actions for each one.

   iv Anticipate possible errors and prevalent error traps for each CRITICAL STEP.

   v Foresee consequences (the worst that could happen) at each CRITICAL STEP.

   vi Evaluate defenses to protect **assets** should error occur; include a review of contingencies and STOP-work criteria.

4 *When*—a pre-job briefing will be conducted for work containing one or more CRITICAL STEPS on the day of the work—just before beginning work.

5 *Where*—the meeting will occur as close to the job's location as possible. It is preferable that the meeting occurs in a quiet location without distraction or interruption.

Usually, in high-risk work environments, the way in which results are achieved are also specified—that is, behavior. If one or more assets are at risk during an operation, you may establish management oversight or specify certain behaviors during performance. For example, you may mandate pre-job briefings and the use of specific **Hu** tools for high-risk jobs. Because of human fallibility, every human endeavor is inherently risky. You cannot always avoid human error, but you can influence its rate of occurrence. Therefore, when working with **hazards** in the workplace, it is important to do the work in certain ways to preserve safety and margin.

If you are sloppy about your expectations, then expect sloppy performance. If people are unsure about what you want, about what is acceptable and unacceptable, people will do their best to give you what they think you want or adhere to the group norm. There can be no real accountability without clear expectations. People do and deliver what your system is set up to create and support.

There can be no real accountability without clear expectations.

What are the true expectations? You may say you want 100 percent procedure compliance at all times. But that may not be what you really get or actually accept. Do you allow discretionary compliance with expectations? Remember, expectations are what managers want—what's written down in procedures, training materials, policies, and management systems—*work-as-imagined*. However, standards of performance usually vary. Standards are what managers and supervisors *accept* or tolerate (by their actions, inactions, what they say, and don't say). What is acceptable becomes *"normal"* practice—*work-as-done*!

Some expectations are more important than others, especially those related to safety. Recall from Chapter 2 that accomplishments can be good or bad: (1) good if the business purposes are achieved safely, or (2) bad if business purposes are not accomplished, or if one or more **assets** suffered injury, loss, or damage in the operational process. Therefore, expectations should address not only what you want to accomplish relative to your business purposes, but also what you want to avoid relative to safety, quality, reliability, and production. Occasionally, deviations from specific safety expectations are not an option—these expectations are sometimes referred to as "red" or "cardinal rules," potentially resulting in termination if violated.

Form clear expectations

In a manner of speaking, expectations are similar to "standard work" used in lean manufacturing applications—a sequence of actions applied to routine work situations.[7] However, expectations, such as **Hu** tools (described in Appendix 3) and other non-technical skills, are not proceduralized, but used when needed, i.e., "skill-of-the-craft." All possible occasions of use cannot be foreseen and should be left to the discretion of the front-line worker. The best expectations adhere to all or most of the following guidelines:[8]

- *Active*—the behavior is an action, something you do, rather than a non-behavior that meets the "dead man rule"; if a dead person can do it, it is not a behavior, such as "stay off the grass."
- *Specific*—taking the form of an explicit behavior or action for an explicit situation.
- *Doable*—the worker is capable of carrying out the behavior in the workplace; reasonable.
- *Observable*—the behavior is visible or countable to anyone when it occurs.
- *Objective*—two or more independent observers would see the same behavior; not subject to interpretation; repeatable.

I encourage you to review the description of self-checking: stop–think–act–review (STAR) in Appendix 3. As you read the description, you should recognize all five elements of a good expectation—you can visualize them in your mind's eye.

Regarding the third guideline, doable, the best expectations are formed in collaboration with those who are expected to adhere to the expectation. Front-line workers know what works and what does not work. The most effective expectations incorporate the input of those who make it happen, taking advantage of feedback provided during a "break-in period" before going live with them, which is discussed in the next section.

Let me make one caveat about adherence to expectations. As a manager, you want predictable, repeatable performance from your people—especially for complex, high-risk work. That's why you develop work processes and write procedures. However, despite your best efforts, something invariably comes up in the workplace that wasn't anticipated by the procedure writer, the designer, or the trainer. Things do not always proceed the way you—managers and designers—assume they will. In such unexpected situations, front-line workers must decide what to do in the absence of time, other experts, and additional resources. This is where RISK-BASED THINKING rises to the forefront of their thinking. People in such work situations must be able to make sensible, conservative decisions that first protect key **assets** from harm. Then, when safety is guaranteed, actions can be taken to achieve production goals. This is commonly referred to as conservative decision-making. See Chapter 4 and Appendix 3 for more detailed descriptions of conservative decision-making.

Do—communicate and train

As a manager, you want to help your people succeed in fulfilling your expectations. Before you can reasonably expect people to perform to your expectations, the organization has to be *aligned* to support the expectation. Otherwise, adherence to the expectation will be short-lived, followed by drift toward a different and unknown standard—whether for good or bad. The preponderance of *local factors* must enable the new behaviors and not create hindrances to their occurrence. Establishing new behaviors requires conducting a *local factor analysis* to ensure the right conditions exist locally and in the system to sustain long-term adherence to the expectation. This method is discussed later in this chapter.

Communicate expectations

Once you form your expectations, you must communicate them to the target population. People must know and agree with the expectation—they must know their duties and responsibilities. To improve the chances that a new expectation captures the hearts and minds of your people, they must believe that the expectation is not only in the organization's best interests—supported with a strong business case—but that it's also achievable (doable) in the workplace. A dialogue must occur with those who will make it happen—usually multiple times via multiple means (or venues). In my experience, I have realized that the following practices improve the acceptance of new expectations:

- *What*—a description of the expectation, including when, where, and support needed (tools, aids, training, coaching, etc.); expectations often include **Hu** tools, which are addressed later in this chapter.
- *Criteria for success*—what is acceptable and unacceptable?
- *How*—the behaviors needed to succeed. For example, what would be seen if someone recorded its occurrence with a video camera?
- *Why*—the business case relative to safety, quality, productivity, customers; contrasting personal costs (what you give up by doing the expectation), benefits (what the person gains in terms of RISK-BASED THINKING when they use it), and risks (of not meeting the expectation).
- *Agreement*—a person's acceptance of the need for and importance of the expectation.
- *7 times/7 ways*—repeating the expectation multiple times, using different ways, media, or venues; including training, which is addressed later in this chapter.
- *Inspection*—the need to observe and measure the occurrence of the expectation on an ongoing basis—people should expect to be observed and not be surprised by it.

When deploying a new expectation, I encourage you to use a break-in period. A break-in period allows those who execute the expectation to try it on for size, so to speak—to see if it will work as advertised. Through candid worker feedback, it gives you time to improve the expectation's usefulness and to identify and correct related organizational/system weaknesses that fail to support it—misalignments. The length of a break-in period varies, depending on the tempo of operations, but generally, two to three months usually suffices. After making adjustments to the expectation and/or related *organizational* and *local factors*, a go-live date is announced at which time everyone is expected to adhere to the expectation going forward.

Training

When people are asked to do something new, they have to believe it is going to help them be successful and that they are able to do it.[9] Without this confidence, people will tend to regress to the "old way." Remember our newly qualified electrician, Kelly back in Chapter 4? Kelly had no experience with the circuit breaker he was asked to operate. He was given no opportunity to practice operating the particular circuit breaker in a training environment before doing it for real—it was simply described during one classroom session. Kelly's supervisor felt procedure compliance was sufficient—which tends to numb independent thinking about safety. Additionally, there was training on **Hu** tools. There was no emphasis to stop when unsure in either training or by his supervisor. Kelly's experience represents a failure of the organization to prepare him adequately for the work he was expected to perform unsupervised. To minimize the occurrence of future knowledge and skill vulnerabilities, managers must assure themselves that front-line workers possess the knowledge, skills, and attitudes necessary to perform their work unsupervised.

The safe and reliable operation of any industrial facility ensures that its personnel, including its managers, possess (1) technical knowledge and skills for their jobs, (2) a thorough understanding of the organization's core technology, (3) an awareness of the key **assets** and their intrinsic risks in the work they perform or oversee, (4) means of exercising positive control (**Hu** tools) of the technology's intrinsic **hazards** and the preservation of **assets**, and (5) a basic appreciation of the business and its economics. Employees must have a healthy respect for the unique **hazards** posed by a company's core technology—such as nuclear power, oil drilling, surgery, flight operations, collecting and storing personal financial information—while at the same time being competent in their roles of supporting its economic operations. In particular, for work involving CRITICAL STEPS, training and qualification requirements must be 100 percent completed before assigning people independent work.

Training creates capability. Training clearly articulates what management wants people to accomplish as well as the new behaviors and attitudes (such as chronic uneasiness) necessary to perform the high-risk tasks. A systematic approach to training (SAT) ensures workers are not only given requisite knowledge and skills for the work, but are inculcated into the values and beliefs of the workgroup and organization. Remember, expertise forms the bedrock for RISK-BASED THINKING. But *beware*, training conventionally applies to singular skills applicable to anticipated work situations. But, to enhance resilience, training that integrates RISK-BASED THINKING must necessarily include education, which emphasizes fundamental understanding of how things work—causes and effects, which can be applied in the workplace to grasp and respond to a variety of novel situations, not addressed by a procedure.[10] Managers who are serious about effective technical training and H&OP apply some form of SAT that incorporates most, if not all, of the following six phases:

1   *Front-end analysis*—define the organizational need—business case. Clarify the organizational opportunity or problem to be solved (safety or production goals not being obtained) by new performance capabilities and capacities for a target population.
2   *Analysis of job-specific training needs*—identify the end performance goals of the training—the accomplishments you're trying to achieve (work outputs), and the tasks and steps (behaviors) that comprise that performance— knowledge, skills, and attitudes. This includes decisions about the nature of the performance such as (1) who performs, (2) under what conditions, (3) within what tolerances (what is acceptable vs. unacceptable), (4) which tasks should be taught, and (5) what is already known by the prospective student.

Ensure you query members of the target population for their thoughts and opinions about their needs.

3 *Design of training activities*—create the blueprint for the training—the instructional strategies that would work best for the target population; format, style, and training activities that best meet learning styles of the trainees; things that will promote learning to ensure learners can actually perform properly and safely on the job.

4 *Development of training*—prepare qualification standards and lesson plans, prepare on-the-job training plans, write programs for computer-based training, prepare student handouts, simulator instructor guides, and examinations, create videos, schedule venues, etc.

5 *Implementation of the training*—conduct training: who, when, where, and how much is needed; carrying out lesson plans, practicing, maintaining course materials, etc.

6 *Evaluation of the training's effectiveness*—collect feedback: (1) reaction of students, (2) students' mastery of knowledge and skills, (3) behaviors on the job, and (4) accomplishment of desired results; determining the validity of the analysis, design, development and implementation. Did the training do what it was intended to do?

In addition to the technical content of a line training program, employees require training on how to anticipate, prevent, and catch human error during work activities—what to avoid. These are part of the mindset of RISK-BASED THINKING and human performance fundamentals that everyone in the organization should be aware of at all times. Training integrates the four elements of RISK-BASED THINKING into training sessions and provides experiences for students to reinforce their practice of RISK-BASED THINKING and chronic uneasiness in the workplace. Simulator scenarios, laboratory training, and case studies of operating experience provide opportunities for responding to unexpected situations—how to adapt to best protect **assets** from harm in uncharted territory for which there are no procedures or guidelines.

When used as a corrective action in response to a problem or **event**, training is legitimate only if there is a knowledge or skill deficiency. Otherwise, training is unnecessary. When training is necessary, too often I have seen managers "wing it" without using SAT for the sake of time. Managers often tell the training manager or instructor to develop a 4-hour training session on "such and such," without considering the basic business need (accomplishments) and the required changes in behavior. Effective training requires a systematic approach that takes into account the conditions of performance along with all the risk-related facets of the work environment. By the way, training is one of the most expensive performance interventions, and when it's necessary, it must be done right.

Once formal training is completed, the new behaviors have to transfer to the workplace. The effectiveness of training depends on supervisory follow-up in the workplace. Supervisors must be prepared to reinforce, coach, and correct, as detailed below. It takes time to inculcate new practices. Occasionally, the supervisor will have to stop work momentarily to coach a worker on a particular technique or expectation. The 5D's "fast training model" works great in the field:[11]

1　Describe it.
2　Demonstrate it.
3　Do it.
4　Debrief it.
5　Do it again (except better this time).

Caution! Be careful with the conduct of on-the-job training (OJT). The more you think you know, the more situational work becomes. Sometimes experienced personnel do not always teach the "book" during OJT. Instead, they may explain "how they do it," i.e., drift. Veteran front-line workers with 20 or more years on the job know the hazards of their processes, but not so much new hires. Be specific about what is taught during OJT, and consider using personnel with three to five years on the job as OJT mentors.[12]

A dull knife doesn't cut. Rusty keys do not open doors. Training is never one-and-done. Your organization's training needs are not static. Technical knowledge and skill requirements, like operational risks, are dynamic and deteriorate over time. Additionally, the physical plant undergoes modifications to equipment, implementation of new technologies, regulatory changes, recent operating experience, procedure revisions, etc. Additionally, human nature continuously works against the very best of training programs—every person's recall of knowledge and skills decays over time, and this decay happens in areas in which the knowledge is rarely needed—but sometimes crucial when it is needed—namely, in emergency situations.

Consequently, technical knowledge and skills need to be refreshed periodically to keep them recallable and usable. Refresher training must be a regular occurrence for front-line workers, where they review not only fundamental technical knowledge and skills but also non-technical knowledge and skills, such as **Hu** fundamentals, **Hu** tools, and other capabilities important to safety, quality, and the business. One of the primary means of detecting the deterioration of knowledge and skills is through direct observation.

Check—observe and inspect

Harkening back to the discussion on drift in Chapter 7, people do not always do what they are expected to do. While on active duty in the U.S. Navy, it was common to hear senior officers say, "You get what you inspect, not what you expect." It takes a relentless pursuit of truth to know what is actually happening in an organization, especially at the sharp end. You have to get out of your office and go into the operating spaces to see firsthand what is really happening—*work-as-done*. Checking involves learning, but learning about what is happening in the workplace. Seeing what people are doing—their choices—is a more reliable indicator of the presence of safety than lagging performance indicators that are published monthly or less often in periodic reports. Field observation, introduced in Chapter 7, is one of a manager's most important learning tools and provides one of the most effective means of SYSTEMS LEARNING.

Consistent adherence to expectations is a hallmark of high-performing organizations. However, to achieve and sustain high performance levels, people, including you, need regular feedback. During field observations, feedback accomplishes two purposes:

- To give workers information about their performance that provides them with the opportunity to change, to more closely track with expectations.
- To provide managers with information about how expectations, working conditions, and management systems related to the work at hand either help or impede the worker.

One of the tenets of high-reliability organizations (HRO) is a "sensitivity to operations."[13] Systems thinking managers, sensitive to operations, routinely walk the operating areas, aware that the behaviors and conditions on the floor are the outcomes of the system. If you regularly visit the shop floor, when things are going well, people will more likely talk with you when things go wrong. Unfortunately, managers are usually unaware of changing demands or of the need for workplace adjustments—even though they are responsible for establishing expectations and criteria for success. If managers and supervisors rarely observe performance, workers may begin to think their performance is not important. The result? They'll begin to take liberties with their work, doing what they think is right, convenient, or expedient, likely unaware of the critical importance of doing their work according to expectations i.e., drift.

Notice that feedback is the core benefit in observation. Here's what Larry Bossidy and Ram Charan said, in their book, *Execution*, about managers who sit in their offices all day.[14]

How good would a sports team be if the coach spent all his time in his office making deals for new players, while delegating actual coaching to an assistant? A coach is effective because he's constantly observing players individually and collectively on the field and in the locker room. That's how he gets to know his players and their capabilities, and how they get firsthand the benefit of his experience, wisdom, and expert feedback. It's no different for a business leader.

In many cases, it is a fiduciary responsibility—a public trust to monitor safety and to address safety issues. The real question is, how are you doing?

"But performance does not mean 'success every time.' Performance is rather a 'batting average.' It will, indeed it must, have room for mistakes and even for failures. What performance has no room for is complacency and low standards."

—Peter Drucker
Author: *Management: Tasks, Responsibilities, Practices* (1973)

Observation and feedback create experiences for the staff. Observation is the only management tool that causes positive change—not only by its use but even by its very existence.[15] Observation affects people's performance because they know that someone could show up to watch what they do. But, when observations happen, will these experiences be perceived as positive and effective, or as negative and unproductive? My professional experience reveals consistently, that when workers see their bosses in the workplace—asking questions and seeking feedback on a regular basis—workers accept the need for feedback and readily share their feelings about the work, especially if their bosses act on the information. Top performers, especially, appreciate that their bosses know them personally and are open to suggestions. When people are treated with dignity and respect, observations become an extremely positive experience for workers, encouraging the flow of information between them, improving your understanding of what works and what doesn't.

According to accountability experts, Roger Connors and Tom Smith in their book, *How Did That Happen?*—managers must monitor operations to:[16]

...assess the condition of how closely key expectations are being fulfilled, to ensure continued [system] alignment, to provide needed support, to reinforce progress, and to promote learning, all in order to bring about the delivery of expected results.

Unless managers make it a point to "go and see for themselves," drift in practices and norms and accumulation of unsafe conditions will remain largely invisible until harm occurs.[17] A manager observer watches both worker and supervisor practices, as well as the conditions affecting (1) their adherence to expectations, (2) the safety of **assets**, (3) the control of **hazards** at **pathways**, (4) compliance with regulations and quality requirements, and (5) productivity and efficiency. When you are in the workplace, have a healthy skepticism. Always believe that you will see actions that are worthy to reinforce, coach, or correct, as well as workplace conditions (*local factors* and defenses) that need to be corrected. I have personally conducted hundreds of observations over my career, and my experience suggests that an effective observation of field activities satisfies the following criteria:

- Both the observer and those observed accept the importance of feedback.
- Managers and supervisors must be able to model expectations before they are allowed to observe and coach worker performance.

- Conduct at least one observation per week for line managers; daily for first-line supervisors; monthly for middle-line managers; and quarterly for senior managers and executives.
- Monitor the conduct of the pre-job briefing (if there is one) and the performance of critical phases of the task or operations (e.g., CRITICAL STEPS).
- Spend at least one hour on the floor observing work in progress for each observation, allowing you to become "part of the woodwork" so to speak. Workers become accustomed to your presence and focus more on the task at hand.
- Identify two or more positive practices (reinforce them on the spot). Do not let great work go unnoticed—which lets people know what is important.
- Identify no more than two opportunities to improve (for an individual worker). Do not let poor work go unnoticed.
- Identify at least one system weakness or faulty defense requiring correction/ improvement.
- Conduct a verbal debriefing with individual(s), achieving agreement on what could be done differently—two-way feedback.

Though the above criteria may be straightforward, effective observation and feedback is not intuitive. Managers and supervisors must be trained and coached on how to do it well. Remember, an observation involves close monitoring of the performance of a work activity, which tends to be intrusive. Consequently, it takes practice to become effective at seeing what needs to be seen and to promote a respectful approach to providing and receiving feedback. To enhance the effectiveness of observation and feedback, it may be worthwhile to provide observers with a tool that describes step-by-step what to do while observing work.

At the heart of observation is collaborative feedback, which requires *engagement* between at least two people. Engagement occurs only when at least two people learn. It is through this face-to-face interaction between manager and worker that learning occurs—identifying and closing performance gaps (see Figure 8.2) as well as discovering system-level weaknesses that contribute to ineffective defenses, land mines, and error traps.[18] And, if field observations are a positive experience for them, workers will look forward to your future visits to their workplace, and be eager to share what they know and make suggestions.

Adjust—give and receive feedback

Training professionals understand that a large majority of learning occurs outside the classroom, on the job. However, without feedback, learning does not occur. Feedback lets people know what is important to you, whether you are giving or receiving it. The kind of feedback you give depends on the behavior you observe and want. The presumption is that you, the observer, know the expectation, explicitly—if you can't recognize desired behavior, then you can't possibly give the proper feedback. The feedback you receive from workers depends on whether you ask for it and then do something with it. Remember, lessons learned are not learned until behavior changes.

Reinforce, coach, correct, or console

There are four basic categories of behavior, and each has its accompanying consequence and feedback technique. As one of the enduring principles of H&OP, every organization is perfectly tuned to get the performance it is getting. A significant reason for this is the consequences people receive for their choices. If at-risk behavior is common, it means managers and supervisors have not made a difference by providing appropriate feedback. So, to be clear on how to use consequences to get desired results, I suggest you use the following four basic approaches dependent on the behavior choices outlined below.

- *The Good*. Desirable behaviors that you want people to repeat—there is no performance gap. *Feedback: reinforce people who practice safe behaviors with positive consequences*. Reinforcement improves the chances that a behavior will be repeated by the individual in the future. People do things because of what happens to them when they do it. By reinforcing people when they do what you want, you'll likely get more of it in the future. You want to use positive reinforcement as often as possible in the first couple of months of implementing a new expectation—otherwise it will stop. It helps to have a plan to (1) reinforce people when they perform according to expectations, and (2) reward people for the results attained using desired behaviors. Generally, without a plan, managers default to negative reinforcement— "do it, or else."[19]
- *The Bad*. Perhaps a bit overstated, but such behavior simply misses the mark—the person's technique is not quite up to your acceptance criteria, or the person made a one-off at-risk choice. *Feedback: coach people to help them improve their technique or more accurately mirror your expectations*. This is an interactive dialogue—a conversation, not a monologue. Coaching identifies a performance gap and facilitates the development of and an agreement on a solution to close the gap. Do not sanction people for one-off (rare) at-risk choices that unknowingly created a risky situation, who were unaware of a safer option, or felt justified in doing what was done. But do it privately— with a positive focus that preserves the person's dignity.
- *The Ugly*. These behaviors are *reckless*, *unsafe*, or *unethical*. They must stop immediately—they must not be allowed to continue. *Feedback: correct or sanction people who exhibit these behaviors—the person has no choice but to discontinue the behavior*. Two or more occurrences of such choices are usually followed with formal counseling. Termination is likely for people deliberately taking substantial and unjustified risks that exposed others to serious risk of harm. (Reckless choices were discussed in Chapter 7 on **event** analysis and will not be explored further here.) Be direct, but do it privately and with respect. Pinpoint the behavior or action to be stopped—in no uncertain terms. If you feel like being merciful, tell the person the consequences of continuing the undesirable behavior (counseling). Then discuss the behavior you want to see in its place. Make sure you get agreement from the person before concluding

the conversation. However, make certain that you use positive reinforcement with the person when he/she exhibits the new behavior (the Good).

- *Human error*. This behavior, by definition, is unintentional—it doesn't fit any of the previous categories. Though it is a performance gap, it is not one directly under the control of the performer. *Feedback*: *console people who commit errors*. This is counter-intuitive for most people. Since workers generally intend to do a good job, making errors on the job are usually disheartening to them. It's hard to manage something people did not intend to do in the first place. People need encouragement in the face of disappointment. Manage the system to accommodate the risk of human error.

When you see it, say it. If you choose not to provide feedback, you are setting a precedent—a new standard. If there is no challenge, you tacitly establish a new behavior that may not be what you want, resulting in an unhealthy drift from your expectations. Your people watch you and what you pay attention to. Seriously, consider the ramifications of providing no feedback for the following behaviors when you are in the workplace:

- *The Good*—"If I don't reinforce the desired behavior, the person may think the behavior is not important and stop doing it."
- *The Bad*—"If I don't coach the deficient behavior, the person will not learn."
- *The Ugly*—"If I don't correct dangerous behavior, others will likely get hurt."
- *Error*—"If I don't console people after an honest mistake, they may feel discouraged, lose self-confidence, and become distracted from their work."

Once you identify needed changes in behavior, what do you do to sustain the change? Managers often mistakenly think, "Just because I said it, it will happen," or "Just because I told them what I needed, they will get it done." You must align the organization to "set the stage" for the performance you want. We now shift our focus to *local factor analysis*, an important management tool for sustaining safe **Hu** in the long term.

Sustaining human performance improvements

Field observation provides opportunities for feedback in the here and now. However, to sustain desired practices, including RISK-BASED THINKING, over the long term by all involved, you must establish stable, local conditions that influence the practices you want and discourage those you don't want—that endure when you are not present. This is where *local factor analysis* comes in handy. Knowledge of which *local factors* are necessary and which ones to eliminate suggests which *organizational factors* need revision.

"Prescription without diagnosis is malpractice," as the saying goes. Before selecting or implementing an intervention or a corrective action for changing behavior, it's important to understand reasons for actual performance and/ or why desired performance is not happening (recall Figures 7.5 and 7.6). In

Chapter 3, I introduced six categories of conditions that influence performance in the workplace, which I refer to as *local factors*.

1   Expectations and feedback.
2   Tools, resources, and jobsite conditions.
3   Incentives and disincentives.
4   Knowledge and skills.
5   Individual capacity.
6   Motives and expectations.

The reasons for performance gaps fall into one or more (usually more) of these categories. Pinpointing reasons for a performance gap is accomplished by conducting an analysis of the factors influencing the behaviors in question—the expected behavior (*work-as-imagined*) and the observed behavior (*work-as-done*). Figure 8.2 provides an illustration of what *local factor analysis* entails.

During the analysis, keep the desired work output (desired accomplishment or deliverable) in focus. Knowing the desired accomplishments suggests the behaviors necessary, but may not be specific enough from a safety perspective. Verify the criteria or specifications that make a work output acceptable (good), rather than unacceptable (bad). It's important not to assume that the expectation, as you "imagined" it, is the best and only way to achieve the desired work product. Your expectation—the "desired" behavior—may not be the best or safest way to achieve what you want to accomplish. Workers will likely be the best source for this insight, if they feel free to give it. So, be open (humble) to revising your expectations.

As you should readily see from the illustration, the immediate work environment either *enables* (aids) or *inhibits* (discourages) behavior. Using Appendices 1 and 2, identify relevant factors that either enable or inhibit the

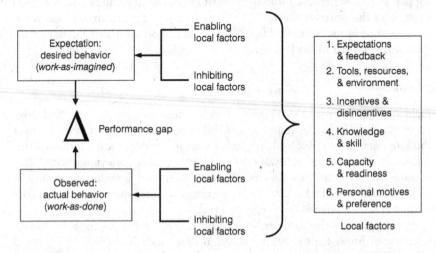

Figure 8.2 Local factor analysis. Diagnosing a performance gap involves identifying the enabling and inhibiting factors for both actual and desired behaviors

target behavior. You'll likely perform the analysis twice, once for the actual/ current behavior (*work-as-done*) and a second time for the expected/desired behavior (*work-as-imagined*).

Looking at the actual behavior first, it should be obvious, since it is actually occurring, that enabling factors prevail over the influence of any inhibiting factors present during work. (Recall the *local factor analysis* method described in **Event** analysis in Chapter 7.) Similarly, analyze the expected behavior to identify both enabling and inhibiting factors. Again, if people are not acting according to expectation, inhibiting factors dominate any enabling factors present in the work context.

Once you identify the most influential *local factors*, you are able to pinpoint specific *organizational factors* that drive the particular *local factor*. For example, training, an organizational function, tends to drive the level of technical knowledge and skill possessed by front-line personnel doing the work (cell 4). To change behavior, you have to change the *local factors*, which, in turn, are created by your system—*organizational factors*.

As you can see in Figure 8.3, the organization drives behavior. This performance model is consistent with the *systems thinking model* and the *alignment model*. All three models illustrate the same fundamental principle—performance in the workplace is a function of the system within which performance occurs. Another way of stating this is that performance is an outcome of various system-related avenues of influence. People are not free agents—over the long term they do what the system encourages or do not do what it discourages.

Until now, I've been addressing expectations generically. Now, I want to shift your attention to a special set of expectations known as **Hu** tools, which have proven to be effective in reducing the occurrence of error across different industries.

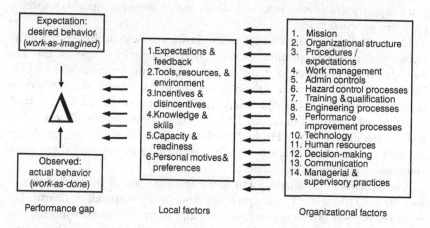

Figure 8.3 Avenues of influence. Sustaining a desired change in workplace **Hu** to more closely resemble your expectations requires the establishment of influences (local factors)—those that enable desired behavior—and eliminating those that inhibit it. (Don't forget to address current unwanted practices.) (Adapted from Figure 1.8 in Watkins, R. and Leigh, D. (Eds.) (2010). *Handbook of Improving Performance in the Workplace*. San Francisco, CA: Pfeiffer.)

Human performance tools

In their book, *Safety at the Sharp End*, Rhona Flin, Paul O'Connor, and Margaret Crichton, describe "non-technical skills" (what I refer to as **Hu** tools) as mental and social skills that complement a worker's technical skills to ensure safe and efficient task performance.[20] Top performers understand that risk awareness is not simply a question of technical expertise (though that definitely helps a lot), but of protecting important **assets** against human fallibility, especially at CRITICAL STEPS and related RIAs. **Hu** tools are a discrete set of behaviors used in the workplace that help users perform their actions more reliably—with less chance of error, focusing attention by "slowing down thinking" to more readily detect and understand important safety-related information. Dr. Rhona Flin and colleagues describe seven skill sets in their book that consistently contribute to safe performance:[21]

- situation awareness
- decision-making
- communications
- teamwork
- leadership
- stress management
- coping with fatigue.

Safe workplace **Hu** depends as much on rigorous use of **Hu** tools skills as on technical knowledge and skills. **Hu** tools help individuals to sustain effectiveness, safety, and even efficiency (in the long run). Appendix 3 provides ten **Hu** tools that have been codified and validated effective in several industries. I encourage you to apply them as is or tailored to work within your organization's specific purposes and constraints.

Hu tools are also known in other industries as seamanship, airmanship, craftsmanship, good operating practice, or best practice. Most of the **Hu** tools described in Appendix 3 are an outgrowth of the practices aboard U.S. Navy nuclear submarines and those advocated in the aviation industry's crew resource management (CRM) training. CRM continues to be practiced by large international airlines.[22] During the 1990s, the U.S. commercial nuclear industry codified the more effective **Hu** tools by documenting what the best performers did that differentiated them from other performers in accomplishing work consistently without error or defect.[23]

The benefits of an effective **Hu** tool encourage one or more elements of RISK-BASED THINKING: (1) anticipate, (2) monitor, (3) respond, or (learn). For example, pre-job briefings emphasize anticipation, whereas post-job reviews focus on learning. Your best **Hu** tools combine all aspects of RISK-BASED THINKING, such as self-checking: *stop* (anticipate and monitor), *think* (anticipate and learn), *act* (respond), and *review* (learn and respond) (see STAR in Appendix 3). In every case, **Hu** tools slow the user down to enable more focused thinking.

The basic purpose of **Hu** tools is to help individuals maintain positive control of their work to avoid at-risk or unsafe behaviors. Intended to control variability and fallibility through a more mindful approach to work, **Hu** tools augment operational procedures, supervision, and the worker's technical knowledge and skills so work can be accomplished safely.[24] Positive control must be focused at CRITICAL STEPS and RIAs to make sure that what is intended to happen is what happens, and that is all that happens. **Hu** tools are not unique to high-hazard situations but are simply good work habits that become particularly important in risky environments, especially where procedures are underspecified and supervision is inadequate. Some **Hu** tools help users recognize when work should stop and when to adapt. Being non-technical in nature, **Hu** tools can be adapted and used in any technical or industrial workplace.

Although **Hu** tools do not guarantee perfect performance, they can greatly reduce the chances of erring when workers use the tools mindfully and faithfully.

Caution! Managers should avoid the temptation to associate the "non-use" or "improper use" of **Hu** tools as a "cause" of **events**. Because of human fallibility, errors continue to occur (albeit at a much lower frequency) despite the regular use of **Hu** tools. Do not "blame" someone for not using a **Hu** tool. Human fallibility applies to the use of **Hu** tools just as to anything else people put their hands to do. Keep in mind that most people want to do a good job.

Time to think vs. time to do

Work environments generally encourage production speed and efficiency in order to maximize output. Overall productivity is maximized if things don't move faster than safety allows. Otherwise, the person or the organization is gambling that mistakes will not happen. There is always a limited amount of time available to do any task, and, given a set time to do any job, the most common variables are time spent thinking and the time spent doing. Although safety often depends on spending adequate time thinking through a task to recognize the risks and what to do to control the risks, people are tempted to reduce thinking time and spend more time doing, especially when time is in short supply. All jobs demand time to perform, which involve thinking and doing. Thus, the time to perform a job can be described generally by the following relationship:

$$T_j \approx T_t + T_d$$

where T_j represents the total time to do a job and T_t and T_d are the components of time devoted to thinking about what is to be done (planning) and time doing what needs to be done (work).

In light of RISK-BASED THINKING, the mindful use of **Hu** tools gives the individual time to consider the importance of his or her task—to think about what will or could happen, what is happening, and what to do if things do not go as expected. The use of **Hu** tools carves out *time to think*, whether working alone or in a group. Just as carpenters "measure twice—cut once," anyone has time to apply the concept of a "double check" to their work.

When faced with time pressure, people are tempted to hurry—to take shortcuts. But true professionals subordinate their moods and take the time necessary to mindfully apply **Hu** tools. Slowing down the work allows the user to better anticipate, perceive, and respond to risks (**assets**, **hazards**, **pathways**, and **touchpoints**). Every **Hu** tool slows down a work process so as to ultimately speed things up by avoiding errors and the delays that accompany unwanted **events**.

> *Warning!* Just as any procedure that becomes highly practiced, **Hu** tools are susceptible to being used mindlessly. You can read a procedure without really understanding what you just read. Or, when using three-part verbal communications, you may find yourself thinking about something else when you get to the third leg of an important communication.

The mindful use of **Hu** tools reduces the chance of human error and provides a "finishing touch" on our grasp of how to reduce the frequency of **events**. Since we can never achieve perfection in avoiding every single error, though, defenses still must be installed to avoid or minimize the damage that results when the inevitable occurs: human error.

Clear **Hu** *tool expectations*

Once a customized set of **Hu** tools have been developed for an organization, managers and supervisors must be capable of demonstrating—modeling—the use of **Hu** tools to others in order to successfully coach, correct, and reinforce their proper use. Similarly, technical training programs should incorporate relevant **Hu** tools to practice and reinforce these skills during hands-on / on-the-job training. During training users can be trained to adapt the tools for specific situations—improving their scalability commensurate with the risk.

> *Warning!* In the attempt to "please their bosses," some users of **Hu** tools could focus too much on the application of the tool instead of the task at hand. This defeats the intent of any **Hu** tool. It has been my experience that occasionally **events** occurred while workers were "distracted" by making sure they performed a **Hu** tool to avoid blame for errors.

Most companies employ competent people, and yet many organizations populated with competent workers can be seriously unsafe. Competence does not necessarily offer positive control against error, and it can even have a negative impact on safety. Safety does not necessarily result from a facility filled with knowledgeable employees because even the most competent people make mistakes. They are, as I have pointed out, human. When an individual relies on his or her personal experience, proficiency, or qualifications during risk-important activities rather than mindfully using **Hu** tools and assumes that his or her experience will help the job to go safely, the attitude has created an unsafe situation. People are not natural rule followers. How many of us have read our vehicle's owner's manual? How many fathers read instructions before assembling toys on Christmas Eve? People tend to focus on efficiency, being as safe as they perceive they need to be. And if past practices have not led to an **event**, an individual worker can become indifferent to the need for care and attention. Top performers, alternatively, routinely and humbly adopt safe practices despite their perception of their own competence. They respect their innate capacity to err.

Accountability and just culture

People are responsible for the choices they make—about what they can control and the risks they choose to take.[25] Too often society thinks of accountability as making restitution—"someone has to pay." Dr. Lucian Leape, a medical physician and professor at Harvard School of Public Health, active in trying to improve the medical system to reduce medical error, said the following during a congressional testimony:

> Approaches that focus on punishing individuals instead of changing systems provide strong incentives for people to report only those errors they cannot hide. Thus, a punitive approach shuts off the information that is needed to identify faulty systems and create safer ones. In a punitive system, no one learns from their mistakes.[26]

Information about what is truly happening in the workplace is essential to identifying and correcting system weaknesses. To foster *learning*—to promote the free flow of information—people must know that they will not be penalized for their actions when they had the best interests of the organization at heart. High-performing, principled organizations do not punish employees for making mistakes when they are trying to do the right thing (see *local rationality* in Chapter 7). James Reason claimed that "Blaming people for their errors—though emotionally satisfying—will have little or no effect on their future fallibility."[27]

But, a "no harm, no foul" approach doesn't work either. There is a time and a place for sanction, especially for reckless and repeated at-risk choices, even when there are no adverse consequences. It's the behavior that must be held accountable. There must be an accounting for decisions to gamble in the pursuit of personal interests. Punishment is reserved for people who ignore safer alternatives.[28] People must know that there are consequences for unsafe behaviors. Without accountability, performance standards deteriorate, schedules slip, managers acquiesce to living with unsafe conditions in deference to production pressures (accumulation), supervisors would spend most of their time in their offices, and front-line workers, be tempted to cut corners, become overconfident, regularly making at-risk choices (drift), or acting recklessly. Consistent, repeatable, high-levels of performance would disappear. Consequently, accountability must be preserved.

Accountability simply means giving people the opportunity to "report" what happened—to tell their story of an occurrence or a situation as they perceived it. A person is accountable when that person is "obligated to report, explain, or justify something; responsible; answerable."[29] But, accountability is predicated on the pre-existence of responsibility. The best explanation of responsibility came from the mouth of Admiral Rickover during one of his congressional hearings:[30]

"Responsibility is a unique concept: it can only reside and inhere in a single individual. You may share it with others, but your portion is not diminished. You may delegate it, but it is still with you. Even if you do not recognize it or admit its presence, you cannot escape it. If responsibility is rightfully yours, no evasion, or ignorance or passing the blame can shift the burden to someone else. Unless you can point your finger at the man who is responsible when something goes wrong, then you have never had anyone really responsible."

—Hyman Rickover

Responsibility only exists when a person clearly understands and accepts the duty to accomplish work according to your expectations, policies, rules, and procedures. Such an "obligation" is usually established by employment. But, as stated previously, without clear expectations, there can be no accountability. Expectations must be clear, understood, and agreed to by your subordinates. But, the "way" accountability is exercised depends on how managers think about people and their systems.

Systems thinking implies that whoever designs the system owns the result— is responsible.[31] Traditionally, the severity of an **event** has been used as the criterion for determining whether punishment or discipline is necessary. When things don't go as expected, managers tend to view individual performance and accountability improperly. Since senior managers devise systems and line managers, supervisors, and workers execute them, individual managers must remember that the management team is "accountable" for the results.

Three views of accountability

There are three perspectives, or views, that affect the way many managers react to human error that triggers an **event**. One is blatantly false, one is an accurate assumption about most people, and one is a basic truth of organizations and related systems.[32]

- *The accountability fallacy*—"When something goes wrong, there's something wrong with the person who 'fouled up.'" Blame the guy who last touched it. He is presumed guilty until proven innocent. To "blame" someone for a mistake is to imply the individual should have perceived the error before taking action, and presumably, he or she voluntarily contributed to the error. If you fix him—make him more conscientious, more careful, or maybe just replace him—the problem will be solved, right? This is myopic, person-centered thinking at its worst. It may temporarily take the heat for a problem away from the folks in charge, but it absolutely will *not* make the operation safer—there is no fundamental change in the system.
- *The accountability assumption*—"People don't come to work to fail; they want to be effective." Most people want to do a good job. Most people readily accept responsibility for their actions and feel real remorse about bad outcomes. These are good people. If adopted as part of your corporate culture, this way of thinking will go a long way toward stimulating feedback. This is not always true for everyone. Occasionally, there are some "bad apples;" people who want to do harm.[33]
- *The accountability truth*—"When something goes wrong, there is something wrong with the system." Surprise! There is something wrong or absent from my system that exposes important **assets** to an unacceptable (uncontrollable) risk of harm—something in the system that neither my staff nor I anticipated. Managing risk within the system is what this book is all about.

Understanding and applying these perspectives will improve your awareness about what actually goes on in your organization and where the real "root" causes lay. If you reject the accountability fallacy, you'll naturally expand the scope of any **event** analysis. Instead of succumbing to hindsight bias and fixating on one small part of the problem—the one that triggered the **event**—you'll "zoom out" for a wider view of the issue—the system. When you make the right assumption about individuals and their rational motives, you'll come to appreciate the value they add to accommodating unanticipated problems. Managers and executives in "just cultures" adopt the accountability assumption and the accountability truth.

Just culture

Learning is more effective when healthy relationships exist and frequent interactions occur between leaders and subordinates.[34] A just culture—how we hold each other accountable—improves communication through greater trust. An atmosphere of trust is cultivated in which people are encouraged, reinforced, and even rewarded, for disclosing safety-related information about workplace practices and conditions. But, trust is further enhanced when there is consistency between what is acceptable and unacceptable behavior.[35]

You cannot blame your way to a better or safer future. Simply blaming or shaming people for errors accomplishes no good and inhibits learning.[36] Blame sets up the organization as well as the individual for repeat poor performance.

No one learns. We all fear embarrassment, losing face, disciplinary action, and even losing our jobs. In a "crime and punishment" work climate, people will not trust their bosses. In businesses where blaming and "CYA" attitudes are allowed to flourish, people game the system jockeying for position, currying favor with bosses, avoiding taking responsibility for problems, hiding their mistakes, and even undermining co-workers. Instead of focusing on the safety of the task at hand, people will tend to give inordinate attention to covering their tracks. Such a climate impairs people's ability to see clearly the adjustments they may have to make to perform safely—much less, report problems. Front-line workers—people doing work at the sharp end of the organization—are the richest source of information about the functioning of your system. You can avoid jeopardizing the flow of this information by understanding and applying one important fundamental principle about the inherent value of people.

"Every human being is intrinsically valuable.
Every human being is equally valuable.
Every human being is exceptionally valuable."

—John Ensor
Pastor

Human beings are not valuable because of their usefulness to serve the desires of others. First, they have intrinsic value just because they are human.[37] Second, no one person's life has more value than another, as acknowledged in first line of the *United States Declaration of Independence*—"We hold these truths to be self-evident, that all men are created equal..." We are equally valuable, though we individually possess different talents and roles. Finally, all persons have exceptional value in that humans have a conscience, are able to reason, know right and wrong, and are not inanimate things. *Relationships are important—principle number 1.* Perceived injustices at work are usually not due to poor performance as much as they are the outcome of bad relationships.[38] There are no laws against treating people with dignity, respect, fairness, and honesty, and everyone wants to be treated this way. Accountability can be exercised without violating people's dignity. The *dignity of persons* must be preserved in any workplace. Violating this principle poses a serious obstacle in achieving open and effective communications with the workforce, ultimately hindering SYSTEMS LEARNING.

People in your organization need to appreciate that they will not be singled out unfairly as the cause of an **event**. When this happens, they will grow to trust that neither you nor the organization are "out to get me" and that any investigation into an **event** is looking for the real truth about how to make life at the company better, safer, more resilient, and more reliable. There is a middle ground between a highly punitive system and a system where no one is accountable (blame free), where both accountability and learning coexist.[39] Accountability is necessarily backward focused, trying to understand the reasons for people's choices, acknowledging miscues, misjudgments, and the harm caused, in light of the system context of their performance. Learning, on the other hand, is proactive. It looks forward to what can be done differently to reduce (1) the potential for error, (2) temptations to make at-risk choices, and (3) the severity of harm.[40] For this to happen, people must feel safe—finding problems before they find you.

When failure occurs in a just culture, instead of asking how an individual failed the organization, managers ask, "How did our organization fail the individual?" What flaws or oversights in work processes, policies, training, procedures, etc. contributed, promoted, or allowed the choice and resulting **event** to occur? Don't think individual accountability is absent—it is very much present. Everyone must tell his or her story, but they must feel safe to reveal it in the necessary detail to learn from it. No one is out to fail. However, since the overwhelming majority of the causes of **events** originate in the organization, management's first reaction to unwanted occurrences should be to question the system. Remember, it takes teamwork and cooperation to suffer **events**. You, the leader, just as the front-line worker, must be willing to give an accounting for weaknesses in the system. Remember the seventh of Admiral Rickover's time-tested industrial principles: "Accept responsibility. Don't pass the buck—admit personal responsibility when things go wrong." If workers see their managers questioning their own ability to get things right, it builds a bridge to help instill right thinking at the operational level. When front-line workers see that no one is above making errors, the "trust meter" goes up. But, to do this takes courage and humility. To get better, to be resilient, the focus must shift from blaming to learning.

Without a just culture, you will not know what is actually happening because communication is inhibited. With a just culture, people find the freedom to concentrate on doing a quality job—without concern for limiting their liability. They make better operational decisions, especially conservative decisions, not afraid to put **assets** in a safe condition in the face of significant production pressures.

How to build a just culture

Just cultures don't just happen. Trust develops gradually as the organization's managers build a social climate that is consistently and reliably fair to everyone in it, acknowledging that no one comes to work to fail—the accountability assumption. Some key success factors for engineering a just culture include:

- Clear expectations—you can't hold anyone accountable to unclear expectations.
- Availability of a fast, easy, and simple means of reporting (format, length, and content of reporting mechanism is not burdensome).
- Encouragement and reinforcement of people for the proactive admission of errors, close calls, and the identification of unsafe conditions.
- Indemnity against disciplinary action for mistakes and one-off at-risk choices.
- Consistent fair application of appropriate sanctions for individuals making reckless choices and others with a history of at-risk choices.
- Reject "no harm, no foul."

- Accountability is administered independent of the severity of an **event**— sanctions are based on behavior choices, not consequences.
- Confidentiality, if desired by the reporter.
- Rapid and meaningful feedback to the reporter about the resolution of the issue reported (something useful is done with the report).

I want to reiterate that using an **event's** consequences to establish accountability diminishes trust. While consequences of an **event** greatly impact people's emotions, disciplinary action should be slow in coming and guided by the individual's behaviors—not by the consequences.[41] Determining whether or not to sanction an individual based solely on an **event's** severity is unjust— and, I believe, immoral. Yet, if people choose to act recklessly, whether harm was realized or not, disciplinary action is justified and necessary. Also, a well-intended but dysfunctional approach, such as a "zero tolerance" policy toward human error, creates frustration and solves nothing—human beings err, and the **events** that ensue are really organizational failures. Avoid policies that attempt the impossible!

You can engineer performance

Real management, especially in high-hazard operations, requires control over what happens (1) to ensure the economic mission is accomplished, and (2) to defend against harm and unwanted outcomes. Throughout this book, I have emphasized that people create safety by the choices they make. To influence people's choices, managers must understand how their systems work and how to leverage them toward safe practices. Dr. Tom Gilbert emphasized this sentiment in his book, *Human Competence: Engineering Worthy Performance*.[42]

> As managers come to see their jobs as largely manipulating the [work] environment in order to achieve greater competence [in their workers], they will become more competent [managers]. Managers who let their employees know what is expected of them, give them adequate guidance to perform well, supply them with the finest tools of their trade, reward them well, and give them useful training will probably have done their jobs well. [Words in brackets added for clarification]

Along with building a just culture, you can engineer performance in the workplace.[43] As described in Chapters 3 and 6, managers can align their organizations (*organizational factors*) to establish workplaces (*local factors*) that influence people's behavior choices at the sharp end to adapt, control, and learn. Mindful managers control those *local factors* that enable desired behaviors and inhibit undesirable choices. Effective system alignment ultimately improves not only safety and resilience, but also productivity and accomplishments. And again, I stress that such alignment depends on managers (1) understanding how their organizations work, and (2) executing sound management of H&OP: plan–do–check–adjust.

Things you can do tomorrow

1 Do you have a "management model?" Discuss this with your direct reports or colleagues—What does it mean for our organization to manage something? What does it look like?

2 In response to the next **event**, console human error, coach one-off at-risk choices, punish recklessness, regardless of outcome.[44] Is the organization's official disciplinary policy consistent with these responses to people's behavior choices?

3 The next time you're observing people do their work, at a convenient moment, ask someone to rate themselves on the task from 1 (poor) to 10 (excellent). When they respond with 7, say, "Excellent, but why aren't you a 1? What do you do well?" After listening to their explanation, reply, "Sounds more like an 8 to me."[45] Then segue the discussion into what it would take to be a 10 from a safety perspective.

4 Use the 5D's fast training model to review a targeted expectation with an individual or work group. Additionally, explore whether the expectation is clearly understood by the individual(s). Ask if he/she agrees with the expectation (why or why not?) and solicit suggestions for improving it (if needed).

5 When work does not go as planned, check on the adequacy of information, staffing, knowledge, skills, tools, procedures, equipment, clarity of expectations, previous coaching. Use Appendices 1 and 2 as a guide.

6 Remove obstacles (hindrances) to desired behavior. Remove negative consequences (punishers) that inhibit an important expectation or rule.

7 During observations, ask hypothetical questions to get conversations about safety going. For example, ask the workers if there is anything slow or uncomfortable about doing the job or task safely.[46]

8 For those *local factors* that inhibit desired performance, look for related *organizational factors* that tend to prevent, delay, inhibit, or divert performance. Look for interactions among various *organizational factors* that create such conditions or unsafe situations.

9 Identify goal conflicts that tempt people to make trade-offs or adapt improperly to accomplish their work.

10 Personally, provide positive feedback (appreciated by the individual) for the occurrence of specific behaviors when they occur. Be specific.

Notes

1 Daniels, A. (1994). *Bringing Out the Best in People: How to Apply the Astonishing Power of Positive Reinforcement*. New York: McGraw-Hill (p.xiii).

2 Rickover, Adm. H.G. (August 1979). Differences Between Naval Reactor and Commercial Nuclear Plants. Comments subsequent to the accident at Three Mile Island. (p.29). Washington, DC: U.S. Navy (NAVSEA 08, Naval Reactors).

3 Reason, J. (1997). *Managing the Risks of Organizational Accidents*. Aldershot: Ashgate (pp.127–128, 223).

4   Walton, M. (1986). *The Deming Management Method*. New York: Perigee (pp. 86–88).

5   The Deming management cycle, based on the scientific method, used the Shewhart cycle: Plan–Do–Check–Act (PDCA). I replaced *Act* with *Adjust*, which seems a more descriptive word to me, if the original plan or progress was insufficient to close the gap. The word adjust suggests managing is an iterative process and aligns with the concept of Risk-Based Thinking.

6   Spector, R. and McCarthy, P. (2012). *The Nordstrom Way to Customer Service Excellence: The Handbook for Becoming the "Nordstrom" of Your Industry*. New York: Wiley (pp.29, 99–101).

7   According to James Womack and Daniel Jones in their book, *Lean Thinking* (2003), "standard work" is a precise description of a production work activity specifying its cycle time, the sequence of tasks, and the minimum inventory of resources on hand necessary to complete the work. It is considered "the best way to get the job done in the amount of time available and how to get the job done right the first time, every time" (p.113).

8   Daniels, A. (1989). *Performance Management*, Atlanta, GA: Performance Management Publications (pp. 135–140).

9   Patterson, K., Grenny, J., Maxfield, D., McMillan, R. and Switzler, A. (2008). *Influencer: The Power to Change Anything*. New York: McGraw-Hill (pp.49–50).

10   McChrystal, S., Collins, T., Silverman, D. and Fussell, C. (2015). *Team of Teams: New Rules of Engagement for a Complex World*. New York: Portfolio (p.153).

11   Connors, R. and Smith, T. (2009). *How Did that Happen: Holding People Accountable for Results the Positive, Principled Way*. New York: Portfolio (pp. 185–186).

12   Marx, D. (2015). *Dave's Subs*. Plano, TX: By Your Side Studios (p.238).

13   *Sensitivity to Operations* means the organization's members understand the importance of its operations—focusing on (1) the planning, preparation, and risk management of production processes as well as safety (anticipation), (2) requiring a high level of situation awareness (monitoring), (3) acting and deploying resources at the appropriate time to protect assets (respond), and then (4) understanding the implications of situations and events—using this information to improve resilience to future events (learning).

14   Bossidy, L. and Charan, R. (2002). *Execution: The Discipline of Getting Things Done*. New York: Crown (pp.24–25).

15   Subalusky, W. (2006). *The Observant Eye: Using It to Understand and Improve Performance*. (self-published at booksurge.com, p.6).

16   Connors, R. and Smith, T. (2009). *How Did That Happen?* New York: Penguin (p.99).

17   Reason, J. (2013). *A Life in Error: From Little Slips to Big Disasters*. Farnham: Ashgate (p.91).

18   I am indebted to Rey Gonzalez for his insight and wisdom regarding the conduct of observations, especially the effectiveness of engagement.

19   Agnew, J. and Daniels, A. (2010). *Safe By Accident?* Atlanta, GA: PMP (pp.9–12).

20   Flin, R., O'Connor, P. and Crichton, M. (2008). *Safety at the Sharp End: A Guide to Non-Technical Skills*, Farnham: Ashgate (p.1). (See also James Reason's reference to "mental skills" in his book, *The Human Contribution* (p.1).)

21   Ibid. (p.1).

22   Ibid. (p.247).

23   Institute of Nuclear Power Operations (2006). *Human Performance Tools for Workers* (INPO 06-002) (p.2). Not available to the general public.

24   Flin, R., O'Connor, P. and Crichton, M. (2008). *Safety at the Sharp End: A Guide to Non-Technical Skills*, Farnham: Ashgate (p.10).

25   Marx, D. (2015). *Dave's Subs: A Novel Story About Workplace Accountability*. Plano, TX: By Your Side Studios (p.139).

26   Patient Safety: Hearings before the U.S. Senate Subcommittee on Labor, Health and Human Services, and Education, and the Committee on Veteran Affairs. Senate.

Volume 146, Number 3 (January 25, 2000) (Testimony of Dr. Lucian Leape). Senate, 106th Cong. 6 (2000) (Testimony of Ira Glasser).

27 Reason, J. (1997). *Managing the Risks of Organizational Accidents*. Aldershot: Ashgate (p.154).

28 Marx, D. (2009). *Whack-a-Mole* (2nd edn). Plano, TX: By Your Side (pp.191–197).

29 Skousen, T. (16 April 2016). Responsibility vs. Accountability. Retrieved 27 May 2017 from www.partnersinleadership.com/insights-publications/responsibility-vs-accountability.

30 Joint Committee on Atomic Energy, Radiation Safety and Regulation, 87th Cong., 1st sess. (Washington, DC: G.P.O., 1961, p. 366).

31 Marx, D. (2009). *Whack-a-Mole: The Price We Pay for Expecting Perfection*. Plano, TX: By Your Side Studios (pp.63, 73).

32 Connors, R. and Smith, T. (2009). *How Did that Happen: Holding People Accountable for Results the Positive, Principled Way*. New York: Portfolio (pp. 17–18).

33 In her 2005 book, *The Sociopath Next Door*, psychologist Martha Stout warned that sociopaths make up about 4 percent of the U.S. population—people without conscience, causing disruptions disproportionate to their numbers.

34 Connors, R. and Smith, T. (2009). *How Did that Happen: Holding People Accountable for Results the Positive, Principled Way*. New York: Portfolio (pp.11, 20–21).

35 Reason, J. (1997). *Managing the Risks of Organizational Accidents*. Aldershot: Ashgate (pp.191–219).

36 Scientists call this "fundamental attribution error"—a pervasive tendency to blame bad outcomes on a person's perceived "inadequacies" rather than attribute the outcomes to situational factors. See Reason, J. (1990). *Human Error*. Melbourne, Australia: Cambridge (p.212).

37 Ensor, J. (2012). *Answering the Call: Saving Innocent Lives, One Woman at a Time*. Peabody, MA: Hendrickson (p.39).

38 Dekker, S. (2007). *Just Culture: Balancing Safety and Accountability*. Aldershot: Ashgate (p.142).

39 Marx, D. (2009). *Whack-a-Mole: The Price We Pay for Expecting Perfection*. Plano, TX: By Your Side Studios (p.116).

40 Dekker, S. (2007). *Just Culture: Balancing Safety and Accountability*, Farnham: Ashgate (p.24).

41 Marx, D. (1998). The Link Between Employee Mishap Culpability and Aviation Safety (p.18). Thesis, Seattle University School of Law. January 31 1998. Retrieved from http://hfskyway.faa.gov.

42 Gilbert, T. (1996). *Human Competence: Engineering Worthy Performance* (Tribute Edition). Washington, DC: ISPI (p.96). Originally published in 1978.

43 Notice in Tom Gilbert's quote that the things managers can do to influence performance are associated with cells 1 through 4 from the *Local Factors* table (see Appendix 1). Cells 5 and 6 are the least "controllable" from an organizational perspective and tend to have the least leverage.

44 Marx, D. (2015). *Dave's Subs*. Plano, TX: By Your Side Studios (p.248).

45 Marsh, T. (2013). *Talking Safety: A User's Guide to World Class Safety Conversation*. Farnham: Gower (p.90).

46 Marsh, T. (2013). *Talking Safety: A User's Guide to World Class Safety Conversation*. Farnham: Gower (p.74).

9   Integrating RISK-BASED THINKING and executing H&OP

> Most attempts at error management are piecemeal rather than planned, reactive rather than proactive, **event**-driven rather than principle-driven.
>
> James Reason[1]

Integration: an act of combining into a whole; organizing the constituent elements of a system into a coordinated, synchronized whole; alignment.

Execution: a disciplined and systematic process of getting things done.

In this chapter I attempt to provide guidance on the implementation of H&OP in your organization. Essentially, you are attempting to manage two things. First and more practically, you want to manage the risk human error poses to your **assets** during operations. Avoiding harm to **assets** is really a control problem—regulating the transfers of energy, movements of mass, and transmissions of information in a manner that adds values instead of extracting value. This is what execution is about—a systems approach to managing risk. Second and more elusive, you must manage meaning. Do people possess a deep-rooted respect for the operation's technology? Do they perceive the risk potential of their moment-by-moment choices? In their work, do they anticipate the creation of **pathways** between intrinsic **hazards** and important **assets**? Are they mindful of impending transfers of energy, mass, and information? Is learning highly valued and rewarded? Do people humbly accept their humanity—in light of their fallibility and innate limitations? Is the dignity of persons preserved? This is what integration is all about—creating meaning through vocabulary and principle.

Remember the seven principles of managing H&OP

I assume that if you have read this far, you will likely want to make some changes in the way you do business regarding **Hu**. However, before making any substantial modifications to your operation, I suggest you and your management team revisit the seven principles of managing H&OP described in the Introduction. As you read through the following principles, you will likely recognize that these concepts were emphasized throughout this book.

1   People have dignity and inherent value as human beings.
2   People are fallible.
3   People do not come to work to fail.
4   Errors are predictable and manageable.
5   Risk is an inherent, dynamic feature in the way an organization operates.
6   Organizations are perfectly tuned to get the results they are getting.
7   The causes of tomorrow's **events** exist today.

By now, the credence of these principles to H&OP should be clear. But what if people who work for you only give lip service to these truths? What happens if they don't really believe them? If so, it is likely there are some deeply held beliefs that conflict with these principles. People act only on what they truly believe. As various operational situations and questions arise, these core beliefs, if adopted into the way you think and do business, will produce consistency in your decision-making and sustain constancy of your responses to **Hu** risks going forward. Without them, your attempts to integrate RISK-BASED THINKING into your operation will be short-lived, and efforts to manage CRITICAL STEPS and SYSTEMS LEARNING will be an expensive waste of time, effort, and money. You and/or members of your management team will tend to work at cross-purposes. Therefore, full acceptance of these principles is essential to sustained success. If you or your organization's management team does not collectively acknowledge these principles, don't start. Building a business case will ensure these preceding principles are mindfully considered before committing your organization to H&OP and RISK-BASED THINKING.

Build a business case

Most efforts to influence business culture fail because those efforts do not link values, beliefs, and new practices directly to the business. People have to see and understand the business case for changing how they work. A sound business case requires that the benefits, costs, and risks, be stated in clear terms, be tangible, and be realistic.[2] People will do what they believe leads to success in their particular part of the organization. Behavior is belief turned into action. If their team, unit, department, or even division is not connected to the business, then they will tend to do their own thing the way they envision success for them, not the organization. Culture change gets real when behaviors and their outcomes are tied directly to accomplishing the business purpose.

The process of developing a business case provides you and your management team with a disciplined method that helps assess the justification and options for pursuing H&OP and RISK-BASED THINKING. An effective business case concisely describes the following:

1   *Current state of human and organizational performance.* This involves the identification of the organization's strengths, weaknesses, opportunities, and threats (SWOT analysis). Appendix 5 provides a basis for conducting a self-assessment.

2   *Objectives and scope.* Relative to the significance (risk) of the issues identified in the self-assessment, clarify the purposes intended to be achieved through the implementation of H&OP and the integration of Risk-Based Thinking. Pinpoint stakeholders and expected long-term outcomes relative to the business. You may want to conduct a "proof-of-concept" pilot project with a particular workgroup to test the fit and effectiveness of H&OP and Risk-Based Thinking.

3   *Alignment.* Strategically, how well do H&OP and Risk-Based Thinking fit the strategic direction of the organization? Conceptually, how do they conflict or support current corporate and organizational initiatives?

4   *Alternatives.* Identify other means of addressing the risk of human error in your high-hazard or high-risk operations; other options that would better fit your technologies, values, and ways of doing business.

5   *Risk analysis:*

   i   *benefits*—safety and financial incentives; potential reduction in "cost of non-quality"; expected outcomes in savings; personal safety, productivity, on-time delivery, and waste; improved analyses of **events;**

   ii *costs*—the expenses of implementing H&OP and Risk-Based Thinking, such as training and related opportunity costs (personnel and financial resources required);

   iii *risks*—downside risks of deploying and *not* deploying H&OP and Risk-Based Thinking; factors that might jeopardize the anticipated benefits or increase costs, including potential unintended consequences.

6   *Recommendations.* Specific suggestions for proceeding with the selected alternative in the near term (next two or three months), mid-term (within the year), and long term (future years); may recommend such things as responsible executive sponsor, project manager, pilot groups, and implementation strategy.

Approval of the business case signifies management's commitment to H&OP, Risk-Based Thinking, the underlying principles, and their willingness to account for its implementation and effectiveness. Implementation of H&OP and the integration of Risk-Based Thinking into operations are considered mandatory for all members of the organization—it is not optional.

A playbook

The majority of this chapter provides general guidelines for deploying H&OP's functions within an organization. There is no secret recipe or formula—all work situations and related organizations are different—having different functions, products, people, processes, values, etc. Therefore, I encourage you to adapt the guidance in this chapter in a springboard fashion to creatively address your organization's unique realities, cultures, and constraints.

In various team sports, coaches develop a playbook for each contest—a strategic plan—that contains the team's strategic goals, plans, and contingencies for achieving victory based on the collective and individual strengths and weaknesses of its players. Coaches develop playbooks to optimize their offensive and defensive capabilities against their opponents. Similarly, H&OP includes both offensive and defensive features. Of the three core functions of H&OP, one is offensive in nature while the other two tend to be defensive. First, CRITICAL STEPS focus on making sure the right things go right first time, every time—an offensive feature. Second, SYSTEMS LEARNING emphasizes defense by identifying and correcting weak or missing defenses that protect **assets** from harm. And last, RISK-BASED THINKING contains both offensive and defensive practices through enhancing flexible responses to changing workplace risks. Your playbook will guide the deployment of H&OP's functions consistent with your organization's unique needs.

To develop your playbook, I encourage you to first conduct a self-assessment of your organization's current practices and operational conditions using the criteria listed in Appendix 5. A self-assessment helps identify strengths, weaknesses, opportunities, and threats associated with your current operations. A SWOT analysis, using Table 9.1, will help you organize the results of the self-assessment—the starting point of your playbook. Additionally, you could expand the self-assessment to conduct a SWOT analysis for each of the seven principles of managing H&OP.

Integration—enabling RISK-BASED THINKING

What if every person in your organization was a risk analyst? What if everyone performed a "source inspection" of their work before he or she passed it on to their customer?[3] RISK-BASED THINKING involves a fundamental transformation in the way people think about and do their work, the way supervisors supervise, the way engineers design their products, the way managers manage, and the way executives plan and direct—all choices made from a deep-rooted respect

Table 9.1 A simple structure for recording the key insights gained from a careful self-assessment of current practices and conditions associated with H&OP

	Strengths	Weaknesses	Opportunities	Threats
1. RISK-BASED THINKING				
2. CRITICAL STEPS				
3. SYSTEMS LEARNING				
4. Observation and feedback				
5. Training and expertise				
6. Integration and execution				

for the operation's technologies. When RISK-BASED THINKING is successfully integrated into the way people do their work, a fundamental change in the company's DNA ultimately occurs (*changing behavior from the inside out—from the heart*). You can change a behavior overnight with wise management of the work environment's *local factors*, but it tends to take years to change a belief—the heart of a person. Take a long-term perspective—there is no rush. Eventually, with your commitment, engagement, leadership, and follow-through, RISK-BASED THINKING will become the foundation for the way you do work and business, despite what you do with H&OP overall.

I suggest you take a decentralized approach to RISK-BASED THINKING, incorporating the four habits of thought (anticipate, monitor, respond, and learn) into key organizational functions. Safety is sustained by enabling anticipation, monitoring, responsiveness, and learning into all facets of operations, and will manifest itself eventually by a reduction in the frequency and severity of **Hu events**. In practical terms, integration merges the four cornerstone habits of thought into all organizational functions and related management systems to:

1   Enable the ability to *adapt* to changing risk conditions in the workplace— to adjust to unanticipated conditions and emergent **pathways** and **touchpoints.**
2   Enable front-line personnel at the sharp end to *control* CRITICAL STEPS during expected and surprise transfers of energy, mass, and information—to add value and avoid injury, damage, and loss.
3   Enable the organization to *learn* about its pitfalls and systems every day— to detect and correct system weaknesses that inhibit the effectiveness of defenses (controls, barriers, and safeguards).

The word *enable* means the organization is aligned to encourage new, desired behaviors founded on the cornerstone practices of RISK-BASED THINKING and to discourage at-risk, unsafe, and reckless behaviors. You'll know when RISK-BASED THINKING has been successfully integrated into your operations when people practice the four habits of thought in various venues as a matter of preference— whether at the sharp or blunt ends of the organization. Everyone becomes a real-time risk analyst.

The integration of RISK-BASED THINKING cannot be accomplished in silos. It takes teamwork, cooperation, and systems thinking. Integration demands hard work and patience, and requires intensive and ongoing engagement[4] with stakeholders, especially with those who do the work at the sharp end. Engaging front-line workers and their supervisors to encourage RISK-BASED THINKING cannot be overemphasized. The information summarized in this chapter should be a useful springboard for creativity and serve as an informative guide for executives and line managers at all echelons of the organization as they work toward integrating RISK-BASED THINKING into their organizations.

Warning! Do not necessarily expect transformational leaders to emerge from the senior management ranks alone. Champions of H&OP can emerge from any level of the organization and form the basis of viral change. Leadership and passion in H&OP may come from mid-level managers, supervisors, and front-line employees. These individuals serve as the sparks for transformational (viral) change within the organization.

There is no intent to create a H&OP program, though you may need such guidance—for consistency—in executing it in the near term. What is intended is to integrate the four cornerstone habits of thought into the existing management systems. Table 9.2 illustrates a structure for reasoning how to incorporate anticipate, monitor, respond, and learn into the various functions of existing management systems and operational processes. Asking yourself how to enable each cornerstone in each function of a process does this.

If there are no intrinsic risks associated with a particular function, then this mental exercise may be neglected or abbreviated. The process of integration is not intended to create an administrative burden for managers and their staff. The process simply provides managers with a systematic way of considering how to facilitate the organized application of RISK-BASED THINKING in a work process or a management system.

As a brief example, I probed the *work execution process* (introduced in Chapter 6) as to how I could weave RISK-BASED THINKING into the performance of workplace jobs. I simply asked, "How can I enable 'anticipate' in the preparation

Table 9.2 A structure for reasoning about how to integrate RISK-BASED THINKING into operational processes

Integration	Operational process		
How do I enable:	Function 1	Function 2	Function 3
Anticipate? (know what to expect)			
Monitor? (know what to pay attention to)			
Respond? (know what to do)			
Learn? (know what happened, what is happening, what to change)			

Table 9.3 An example of how RISK-BASED THINKING can be integrated into the work execution process

Integration	Work execution process		
How do I enable:	1. Preparation	2. Execution	3. Learning
Anticipate? (know what to expect)	**Review** the task to identify what is to be accomplished (outputs) and what to avoid (harm) **Ask** "What if…?" and how you could be surprised	**Look ahead** to all **pathways**: • Transfers of energy • Movements of mass • Transmissions of information	**Foresee** impact on safety and reliability of future tasks, if nothing changes
Monitor? (know what to pay attention to)	**Review** work procedures that describe the desired accomplishments and expectations **Identify** CRITICAL STEPS and related RIAs **Identify** uncertainty around CRITICAL STEPS	**Concentrate** on: • **Pathways** and TOUCHPOINTS— changes in state of assets, hazards, and their controls • Error traps at CRITICAL STEPS and RIAs • CRITICAL STEPS / RIAs • Critical parameters and limits	**Identify:** • Differences between *work-as-done* and *work-as-planned* • Surprises, errors, and recurring adjustments
Respond? (know what to do)	**Conduct** pre-job briefing just prior to commencing work **Decide** how to exercise positive control of accomplishments and avoid losses of control **Stage** resources **Identify** sources of expertise and communication channels	**Apply** as needed: • **Hu** tools • Hold points • Contingencies • **STOP**-work criteria / timeouts • Conservative decisions • Adjustments	**Conduct** a post-job review **Report** serious differences between *work-as-done* and *work-as-planned*
Learn? (know what happened, what is happening, what to change)	**Recall** relevant operating and personal experience **Ask** "What if …?"	**Be mindful:** • Chronic uneasiness • **Ask** "What if…?" • Observation and feedback	**Identify** needed changes at: • System level • Personal level

stage of the *work execution process*?" I followed that question by asking, "How can I enable 'monitor' in the preparation stage of the *work execution process*?" (see Table 9.3). You don't have to complete one function before addressing the next function. You don't have to review one cornerstone before tackling the next cornerstone. There are no rules as to the sequence. The table structure simply gives you a way to keep track of what has or has not been considered.

Do not think of RISK-BASED THINKING as a program or checklist to be added to your current repertoire of management systems, procedures, and business processes. It's not a project or initiative with starting and stopping points. A program is like a security blanket—it creates a warm, fuzzy feeling, which tends to promote complacency. When people think they are in compliance with all the requirements of a checklist, they tend to stop thinking about safety and shift their focus to the "real business of the organization." Managers are particularly susceptible to this error, and they tend to shift their attention almost exclusively to production activities when the most recently completed audit reports that everything is "in compliance." Managers think "we're safe," and safety drops from the forefront of their thinking, forgetting that safety and resilience must be performed in concert with production activities. Instead, consider RISK-BASED THINKING as a way of thinking and doing business that overlays your current commercial processes and systems.

Execution—deploying H&OP

Execution is about managing—getting from here to there. You have to know where you are and where you want to be in relation to the desired endpoint. Then you develop a plan that will close the gap, followed by execution of the plan. You monitor the plan's execution on a regular periodicity to identify adjustments along the way.

H&OP is *all* about managing risk—this is the vision of the future of H&OP when it is fully deployed. At the peril of seeming overly simplistic, in essence, H&OP involves three essential risk-management processes:

- Front-line workers and supervisors mindfully and rigorously exercise positive control of CRITICAL STEPS and related RIAs during operations.
- Executives and managers relentlessly pursue SYSTEMS LEARNING and follow through with an effective CA/PA process to ensure robust and resilient defenses are in place.
- Everyone adapts to changing risk conditions as needed to protect **assets** from harm—possessing a chronic uneasiness toward all facets of operation and practicing RISK-BASED THINKING.

These three elements—the first three building blocks of H&OP—form the vision of the change—where you want to be. See Figure 9.1. Taken separately, each of the six building blocks are insufficient. But, collectively, these strategies are effective at realizing a reduction in both the frequency and severity of **Hu events**.

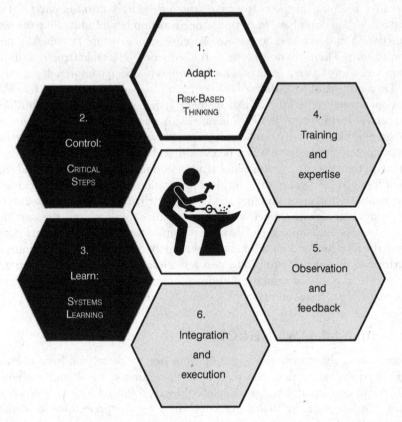

Figure 9.1 The building blocks of H&OP. This model of human and organizational performance accentuates the more important points of leverage in managing the risk of human error in operations

Warning! Do not set out to change the culture. Culture change is not the goal, though it is a desirable outcome.[5] The real emphasis is on managing the risk—behavior choices at **touchpoints** and **pathways** between **assets** and their **hazards**—the application of RISK-BASED THINKING, and verifying the presence and integrity of necessary defenses. Changes in values, beliefs, assumptions, and norms will emerge as people experience success with H&OP and RISK-BASED THINKING. Focus on people's recognition and responses—behavior choices—to high-risk work situations (anticipate, monitor, respond, and learn).

Deploying or executing H&OP involves four stages where the last one is really ongoing:

1   *Exploration.* The organization assesses its needs—it identifies the most important and most frequent operational and **Hu** risks it faces, including a review of its **event** history, and considers adopting specific practices and tools relevant to the specific **Hu** risks. As mentioned earlier, building a business case for H&OP will help clarify your business needs related to human performance. Again, if you cannot buy into the principles of H&OP, I suggest you stop at this point and cut your costs. H&OP is not for you.

2   *Preparation.* Plan–do–check–adjust. Clarify the future state. In light of the desired business results, the organization pinpoints the new behaviors (and unwanted behaviors) and the relevant *local factors* and their related *organizational factors* to sustain the new behaviors. Managers set aside resources necessary to support needed organizational changes that change *local factors* and enable the new behaviors, such as staffing, training, space, equipment, organizational supports, and new operating policies and procedures (organizational changes). Conduct initial training on H&OP for managers and supervisors, followed by training for front-line workers.

3   *Pilot implementation.* Conduct a relatively small-scale, "proof of concept" pilot project where a specific organizational unit at an operations level commits to implement H&OP and integrates RISK-BASED THINKING into its operation. This minimizes your investment risk. Implementation (all six building blocks) should follow the execution activities described below. Implementation should target the most significant and most frequently occurring operational risks. Achieve small wins early. This stage is characterized by frequent feedback and problem-solving at both the sharp and blunt ends of the organization.

4   *Full implementation.* Following the execution activities described below, the organization institutionalizes H&OP into its overall management system and verifies they are functioning effectively to achieve desired business outcomes. The organization matures in its understanding and application of H&OP and its principles. Managers become relentless in identifying and resolving problems, and the organization internalizes RISK-BASED THINKING as the expected way of doing work and business. Eventually, executives expand the use of H&OP and RISK-BASED THINKING to other high-risk operational units of the organization.

Ballpark sequence of execution activities

In light of the preceding section, the following activities will be helpful in the development of your playbook. It is not intended that the sequence be deployed in a literal and linear fashion, but it should be rolled out flexibly and, sometimes, iteratively. Which activities precede others should be apparent. Almost any sequence keeps your effort in the "ballpark." Based on unique operational and organizational circumstances, some of the following items may be performed earlier or later, depending on need or urgency. Additionally, minimize bureaucracy (creating programs), which tends to promote a compliance mentality toward something that is dynamic and uncertain—human performance.

1   *Gain sponsorship from the board and senior management,* if not already done. Develop and review the business case for implementing the building blocks of H&OP.

2   *Familiarize the organization's senior management team with H&OP's strategic approach, building blocks, and core principles.* Ensure the business case for and the risks of **Hu** are clearly understood, that senior leaders commit to H&OP's first principles, systems thinking, and the four habits of thought espoused by RISK-BASED THINKING, and that they recognize the need for resources before embarking on the initiative.

3   *Develop a positive vision of H&OP and RISK-BASED THINKING when it is fully integrated into operations.* Something from each building block is occurring. What would be happening daily? It may be helpful to communicate the initiative in a formal policy statement about the management of **Hu**.

4   *Familiarize line managers and supervisors on the principles and practices of RISK-BASED THINKING and H&OP.* Similar to the introductory seminar conducted for the senior management team, the introductory sessions clearly pinpoint the new ways of thinking and new espoused values to protect **assets**. Verify they understand that H&OP is part of daily core business (cannot be delegated) and what their role is in integrating and executing H&OP operationally.

5   *Conduct an assessment of current practices* (to know where you are). Compare current practices and conditions against the self-assessment benchmarks listed in Appendix 5. A SWOT analysis is very helpful in this effort. Develop action plans to address the more serious vulnerabilities.

6   *Develop a change management playbook.* The playbook should be oriented toward institutionalizing the core building blocks of H&OP. Coordinate the elements of the H&OP initiative with other change initiatives. Resolve cross-purposes, vocabulary, and duplication of effort.

7   *Identify and control the most important, most frequent operational risks.* Identify high-risk, high-frequency work activities, relevant to key **assets** of the organization, and identify their recurring CRITICAL STEPS, related RIAs, and needed controls (CRITICAL STEP MAPPING). Start small and go slowly. It may be worthwhile in the near term to focus only on the "product" as the key **asset** to protect. Also, run a pilot, restricting the scope of the initiative to one or two specific operational units to create "run time" and early wins.

8   *Pinpoint new behaviors* (non-negotiable expectations). Develop these in collaboration with populations expected to adopt the new behaviors. Conduct a *local factor analysis* for each new behavior and work group. For

example, pre-job briefings (new behavior) by supervisors and front-line workers (work groups) or field observations for managers.

9   *Conduct H&OP training*. Use a systematic approach to training. Develop and conduct training on expectations and **Hu** fundamentals for line managers, first-line managers or supervisors, and front-line workers. Everyone needs to understand the basic principles and concepts of H&OP and RISK-BASED THINKING to turn around old ideas about human error.

10 *Conduct field observations and feedback*. Using a systematic approach to training, develop and conduct observation and feedback training. Use paired observations regularly to give managers and supervisors feedback on their effectiveness in conducting observations and feedback. Managers and supervisors must be able to reinforce, coach, correct, and console people when appropriate.

11 *Revise the* **event** *analysis processes* ("root cause analysis"). Eliminate blame. Consider incorporating "local rationality analysis," "local factors analysis," and "behavior choice assessment" tools. Coach analysts and managers as needed to assist them in understanding and applying systems thinking to **event** analysis.

12 *Develop H&OP subject-matter experts*. SMEs serve as facilitators and mentors for line managers and the rest of the organization until line managers develop proficiency and confidence in managing **Hu**.

13 *Optimize SYSTEMS LEARNING*. Assess the effectiveness of SYSTEMS LEARNING methods already in place, and incorporate additional methods (as needed) to address weaknesses in learning. Incorporate leading and proactive safety indicators as well as the lagging, reactive measures. Institute a CA/PA management system of review and accountability for promptly identifying and resolving system weaknesses that could affect the integrity and robustness of defenses associated with high-risk, high-frequency work activities (noted above). SYSTEMS LEARNING should not be perceived as optional. Insist on managerial accountability to encourage follow-through. Reviews of individual managers should occur at least every month for the first six months of the initiative, relaxing the periodic reviews to every 90 days thereafter.

14 *Identify and develop relevant* **Hu** *tools*. If needed develop methods of positive control of CRITICAL STEPS and related Risk-Important Actions (see items 8 and 10). Solicit input from the target population on the **Hu** tools (what, when, where, and how). Again, using a systematic approach to training, develop and conduct hands-on training on specific **Hu** tools. Training on **Hu** tools incorporates multiple opportunities to practice them and to receive feedback. Integrate **Hu** tools into technical training activities.

15 *Develop a robust reporting system*. Regularly collect and resolve feedback from front-line workers about serious and repetitive differences between *work-as-done* and *work-as-imagined*. Make it fast, simple, and easy to use. Consider confidentiality and anonymity features to encourage reporting.

Success factors

Executing H&OP and integrating RISK-BASED THINKING takes time and resources. If you want to get it right the first time with minimal sidetracks, incorporate the factors listed below into your rollout plan. People must clearly recognize that safety goals as well as business results will not be met with old ways of thinking and acting. Again, managers must (1) clearly define the business case, the operational problem with safety and resilience (the Why), and (2) specify new behaviors—expectations (the What). People must understand the value of the change to the business and to them personally (What's in it for me?).

My experience bears out the fact that front-line workers readily accept H&OP and RISK-BASED THINKING. The greatest hindrance to change traditionally has been at the management level because H&OP and RISK-BASED THINKING involves (1) spending more time on the shop floor, (2) giving up some amount of control to those at the sharp end, and (3) increasing their accountability for improvements in safety and resilience. Managers need to conduct field observations, seek and give feedback, be accountable for SYSTEMS LEARNING, commit the resources to technical training, and be willing to treat front-line workers as agents of the organization—allowing them to make conservative decisions in the workplace. Until enlightened by systems thinking, line managers tend to assume that the changes associated with H&OP focus on the front-line worker because of all the talk around human error. Here are the more important success factors managers and leaders should consider when rolling out H&OP:

1   *Assign a senior executive champion for H&OP.* This is more than resource sponsorship. A senior manager puts on the mantle and guardianship of the operating philosophy until it is thoroughly embedded in the culture. As a first step in transformational leadership, the executive champion believes in the credibility of H&OP and RISK-BASED THINKING acting as their cheerleader.
2   *Communicate a compelling vision.* Clarify the business case—the Why. Characterize the operational problem to be solved, preferably supported with data (safety, economic, political, social, and personal reasons). Accentuate survival anxiety—pinpoint dissatisfaction with current practices and conditions. Describe H&OP's costs, its benefits, and, of course, the risks of not doing it.
3   *Use a communication plan.* A communication plan should not only consider formal methods and forums, but also provide guidance for the frequent informal interactions between managers and subordinates. Decide on and emphasize the vocabulary you want to use for corporate-wide consistency. For terminology ideas, refer to the Glossary.
4   *Start small.* The how-do-you-eat-an-elephant principle applies to changing a corporate lifestyle. Conduct a pilot, "proof of concept" project with a

specific operational organizational unit. Focus on bite-sized but critical changes first and celebrate—even your small successes. Start with a small set of non-negotiable behaviors (expectations)—not optional.

5   *Start where you are.* Assess (1) current H&OP management practices, (2) approaches to individual **Hu**, training, **event** analyses, (3) current workplace practices and conditions, (4) recurring high-risk operations, and (5) current Systems Learning processes. Then, you'll know where to target limited resources for initial changes going forward.

6   *Go slowly.* This is a journey. Behaviors can be changed overnight, but changes from the heart can possibly take years.

7   *Target high-risk operations.* Identify Critical Steps and related RIAs for high-risk operations. Consider using Critical Step Mapping for your frequent, high-risk operations. As with isolating Critical Steps, determine which operations and activities are your most hazardous work processes and then begin managing the risk there.

8   *Identify internal champions of H&OP.* Identify volunteers, who are held in high regard by their peers, to serve as subject matter experts (SMEs) positive role models of H&OP and Risk-Based Thinking. Create communication networks/forums for internal champions to facilitate discussion of issues and the exchange of ideas.

9   *Incorporate Risk-Based Thinking at all levels.* H&OP should be reflected in business plans, and Risk-Based Thinking should be consistently applied from the shop floor to the executive boardroom and across divisional boundaries.

10 *Spend time in the workplace.* Managers must see, firsthand, how work is actually done and how their management systems, policies, plans, etc. support or inhibit work.

11 *Collaborate with the sharp end.* Integration requires intensive and ongoing engagement with stakeholders—especially with those who do the work. Encourage viral change through formal as well as informal leaders. Clarify the role of front-line workers as agents for the organization as protectors of **assets**. Give workers input into the means of addressing operational issues and in developing new behaviors (expectations). Buy-in is achieved more quickly when the target populations have a sense of control of their work.

12 *Develop and communicate clear expectations and create feedback opportunities.* Communicate clearly and frequently about the new expectations—new behaviors and work outputs. Two-way feedback occurs quickly and often. Open channels of communication are readily available, preferably through face-to-face interactions in the workplace. Managers must be able to model new expectations—to reinforce, coach, and correct. Always follow through on feedback from front-line workers.

13 *Use a rewards and reinforcement plan.* Reinforce behaviors and reward results. Create reinforcement plans for those associated with the control of Critical Steps and RIAs. People tend to repeat what is reinforced. Managers should

refer to these plans during field observations. Regularly celebrate wins and accomplishments, pinpointing the behaviors that achieved the safe outcomes.

14 *Schedule break-in periods for new behaviors*. Give the target populations time to practice new expectations and **Hu** tools in the workplace. Solicit feedback from users on the expectations to ensure they are doable and effective, and then set a "go-live" date.

15 *Align the organization*. Align the organization's management systems and leadership practices to support new expectations—desired behaviors—by adapting *local factors*, consistent with the organization's business goals.

16 *Adjust schedules—be flexible*. H&OP and RISK-BASED THINKING take time. Create time and capacity in the organization's work schedule for pre-job briefings, for use of **Hu** tools, for post-job reviews, and for continuous improvement.

17 *Make H&OP mandatory, NOT optional*. H&OP requires a commitment by company leadership and management at all levels. Hold managers accountable through regular checks (every 60 to 90 days) to the plan and for consistency with H&OP principles and RISK-BASED THINKING cornerstones.

18 *Manage H&OP*. As with any business operation, implement H&OP and RISK-BASED THINKING using some form of plan–do–check–adjust. Verify that CRITICAL STEPS and related RIAs are identified and controlled. Verify that SYSTEMS LEARNING is occurring and time-at-risk is minimized.

19 *Emphasize the use of positive leading indicators*. Identify leading indicators that increase when safety and resilience improve, such as the number of and time conducting observations. Don't rely solely on negative indicators, such as the number of events or number of operator work-arounds. Managers should determine what needs to be measured in order to evaluate progress and effectiveness of H&OP—e.g., trending, leading, and lagging indicators.

20 *Create and sustain healthy relationships*. Espouse the dignity of persons. This is key to making changes that last.

21 Embrace the vocal skeptic—a converted skeptic is worth 100 disciplined followers!!!

Promoting RISK-BASED THINKING and chronic uneasiness

A strong safety culture requires leaders to actively encourage and reward uneasiness and promote safety behavior choices. A "strong" safety culture is one in which people throughout the organization—not just a few individuals—make safe choices that protect assets from harm. Most of the time harm is avoided by following procedures, policies, and expectations—the rules. But sometimes, following the letter of the law may not be the safe thing to do. A strong safety culture allows a well-trained and qualified workforce to adjust when and where necessary.

Creating a strong safety culture is not the goal. It is behavior choices. Changing behavior requires persistence on the part of every manager and supervisor. It means spending time in the workplace on a regular basis to see firsthand the practices of RISK-BASED THINKING. Safety and resilience demand knowledgeable and energetic leaders ever insistent on high standards. Success in learning and promoting a conservative approach to work requires managers to engage their workers. But to be successful, leaders, from the boardroom to the shop floor, must model anticipation, monitoring, responsiveness, and learning with their immediate subordinates. Together, RISK-BASED THINKING and chronic uneasiness serve as important ways of thinking and feeling about intrinsicly risky operations, enhancing people's adaptive capacity to ensure what absolutely has to go right indeed goes right.

If you want to transform the way people think, consider your vocabulary. Karl Weick, co-author of *Managing the Unexpected*, said, "If you want to change the way you think, change the words you use." This is the quote that precedes the Glossary. For RISK-BASED THINKING to become habits of thought for the organization, managers will necessarily have to talk about these practices and H&OP concepts explicitly, as often as possible, and in as many venues as possible. I strongly suggest that executives, managers, as well as line supervisors understand and use the language of H&OP and RISK-BASED THINKING regularly, especially during field observations and operational meetings. To aid this transition, I suggest line managers serve as the training instructors for the H&OP training of their subordinate personnel. That way you and your colleagues will be forced to become more than one question deep in your understanding of H&OP and RISK-BASED THINKING.

Communication plan—for management

When safety and production goals conflict, what choice will people on the shop floor make? Safety should be the default response. But this is not always the case. Why? The answer is related to the answers to the following questions:[6]

- What do the supervisors, managers, and executives talk about the most?
- What do bosses get mad about?
- What gets measured?
- What do managers pay attention to regularly?
- What actions and outcomes are praised and rewarded?
- What criteria are used to calculate bonuses?

In a high-reliability, resilient organization, the choice to make is always clear, especially when safety boundaries of key **assets** are approached or exceeded. However, for some, less-mature organizations the choice may not be so apparent.

Effective communication doesn't just happen—it follows a plan, not just good wishes that people will talk openly about safety problems. Implementing H&OP and RISK-BASED THINKING requires frequent communication of the desired end state, its principles, and expectations, using the vocabulary. Executives and managers must be sensitive to informal interactions where a careless, unthinking word or action could convey an unintended message. Neglecting to express an appropriate comment when needed may even offer tacit approval of a work practice or condition that should not be endorsed. Through a combination of formal and informal media channels and forums, a communication plan should generally adhere to the following structure and practices:

1   *Purpose.* Clarify the business case, problem(s) to be solved, and goals of the message(s).
2   *Message.* Summarize the key points of the message(s). Think about what *not* to say and how to react to **events**. Incorporate the vocabulary of H&OP (see the Glossary).
3   *Audience.* Identify target audiences (work groups). Focus on influencing key role models—informal leaders—rather than those who habitually complain.
4   *Forums.* Identify a variety of settings in which managers, supervisors, and workers interact, such as production meetings, shop meetings, classrooms, field observations, pre-job briefings, and post-job reviews, even passageways.
5   *Media.* Develop diverse communication media such as newsletters, weekly bulletins, intranet, e-mail messages, closed-circuit television, and posters.
6   *Delivery.* Take frequent opportunities to convey the message(s) when situations present themselves, both planned and extemporaneously.
7   *Checks.* Validate that the right message is being received and understood by the target audience. Ask questions to confirm people's agreement with principles and practices.

"…what I ask questions about sends clear signals to my audience about my priorities, values, and beliefs."

—Edgar Schein
Author: *Organizational Culture and Leadership* (2004)

Your communication plan is essentially an internal marketing plan through which management promotes RISK-BASED THINKING and H&OP. The plan helps you take advantage of both formal and informal occasions to regularly promote H&OP's business case—its vision for safety, its values, priorities, principles, and expectations.

Consistent, clear, and honest messages are important for trusting relationships. It is a vital *local factor* for open reporting of issues, surprises, and mistakes on the job. If people do not understand what you want—what success looks like—the transition to RISK-BASED THINKING will falter.

Finally

There are no silver bullets, but what I have attempted to do is to sharpen your focus on what is to be managed. All work situations and related organizational units are different—with different functions, products, people, processes, values, etc. Therefore, I encourage you to adapt the guidance contained in this chapter (as well as the entire book) in a springboard fashion to address the unique realities, cultures, and constraints of your particular organization. Tailor it to make it work for you and your organization.

Progress will be bumpy in the first year or so as people learn new ways of thinking and doing work. Resistance is likely—as people experience some forms of anxiety in adopting new behaviors that conflict with old habits that are still positively reinforced. This is why positive reinforcement early and often by managers is so important, especially during the first six months or so.[7] Learning to think differently is always difficult. Consider the progress in H&OP and RISK-BASED THINKING as continuous and evolutionary—a journey, not a destination.

Things you can do tomorrow

Actually, many of the following items will likely take some time to do. Don't hurry.

1 Conduct a self-assessment of current H&OP management practices using the criteria listed in Appendix 5.
2 Using your organization's business planning process, develop a first draft business case for H&OP. Talk through it with your management team, identifying needed revisions. Solicit praise, doubts, and reservations about H&OP.
3 Discuss with your colleagues or management team what success would look like in two to three years. What new behaviors would you see happening among front-line workers / supervisors / line managers / executives? (See Table 8.1).
4 Using the building blocks of H&OP, discuss with your colleagues or management team what you are currently doing well and not doing well in each building block. What do you need to do differently?
5 In collaboration with your management colleagues or management team, prepare a playbook to guide your implementation of H&OP for the high-hazard operations of your organization. Identify other initiatives that H&OP may work at cross-purposes with. Consider inviting senior front-line personnel to participate.
6 Include frequent, brief articles in company newsletters that draw attention to successful applications of RISK-BASED THINKING and H&OP. Highlight high-performing persons, especially at the sharp end, who would serve as H&OP SMEs and RISK-BASED THINKING champions and role models.
7 Develop a communication plan that not only addresses formal organization-wide messaging, but also informal one-on-one or small group interactions.
8 Develop a reward and reinforcement plan for targeted work groups and new behaviors.
9 Extend the emphasis on RISK-BASED THINKING and H&OP to other sectors or units of the company.

Notes

1 Reason, J. (1998). *Managing the Risks of Organizational Accidents*. Aldershot: Ashgate (p.126). ·

2 Center for Chemical Process Safety (2007). *Guidelines for Risk Based Process Safety*. Hoboken, NJ: John Wiley (p.651).

3 Traditionally, a source inspection is an examination of the finished product's conformance to quality requirements *before* shipment to a customer. A worker can similarly check that 100% of his/her work meets specifications—improving their ability to detect mistakes and enable corrections before nonconformities are generated.

4 Engagement focuses on two-way learning. It involves face-to-face communication between leader and subordinate, supervisor and operator, manager and supervisor, director and manager, executive and director, CEO and executive. Engagement does not happen until at least two people (in a one-on-one interaction) learn.

5 Schein, E. (2004). *Organizational Culture and Leadership*. San Francisco, CA: Wiley (p.319).

6 Schein, E. (2004). *Organizational Culture and Leadership* (3rd edn). San Francisco, CA: Jossey-Bass (p.246).

7 See *Performance Management* (5th edn) (2014) by Dr. Aubrey Daniels and Jon Bailey.

Epilogue
Live long and prosper

The greeting, "Live long and prosper," was popularized by actor Leonard Nimoy as the character Mr. Spock in the 1960s–1970s-television series, *Star Trek*. Actually, the greeting is a variation of a Jewish blessing recorded in Deuteronomy 5:33 of the Holy Bible (NIV).

To maintain safe, reliable, and resilient operations, an organization's leadership must have the will to communicate—fearlessly. By cultivating an atmosphere of trust, achieved through honesty, fairness, and respect, candid conversations about safety between employees and managers can become commonplace. In such an environment, everyone willingly accepts responsibility for their choices and seeks assistance by admitting to and learning from their mistakes, knowing it's in the best interest of the organization's prosperity and the safety and health of its people. This requires courage, humility, and an unwavering dedication to the dignity of persons.

A corporate culture in which RISK-BASED THINKING is integrated into all aspects of work affects everything about how a company does business. What else is needed if everyone in your organization practices RISK-BASED THINKING—everyone anticipates, monitors, responds, and learns? People in a strong safety-culture start work knowing they are expected to apply RISK-BASED THINKING to whatever job or project they are assigned.

In Chapter 1 you became familiar with my story. You could characterize some of my life's experiences as lucky. To my surprise, many highly respected scientists believe in luck. I for one do not think that everything that happens to us is pure happenstance. I believe that reasoning, human error, and "luck" are all part of a sovereign design. Physics and faith are intertwined. I believe I should do all I can to proactively achieve my goals, avoiding harm in doing so. But, at the same time, I depend on a Creator, who makes all things work together for good. Is it so hard to comprehend?

Above all else, I am passionate about safety because of the innate sacredness and dignity of human life. Nothing is greater than protecting this gift each of us received from a Creator, who had the vision to make a being as wonderful and yet so mysteriously fallible as humankind. I share the awe expressed by the psalmist when he inquired:

⁴   What is man that You are mindful of him,
    And the son of man that You visit him?
⁵   For You have made him a little lower than the angels,
    And You have crowned him with glory and honor.
⁶   You have made him to have *dominion* over the works of Your hands;
    You have put all things under his feet.

—*The Holy Bible*, Psalms 8:4–6
(New King James Version)

My earnest prayer for you is that you will exercise dominion over the work of your hands using a risk-based approach so that you and yours will *live long and prosper*!

Appendix 1
Generic local factors

Table A1.1 Generic local factors

1. Expectations and feedback	2. Tools, resources, and job-site conditions	3. Incentives and disincentives
a. Expectations specify what is to be accomplished (work outputs), criteria for success, behaviors, and under what conditions.	a. Adequate time and flexibility are allotted to accomplish expectations without compromising safety and reliability.	a. People are treated fairly, honestly, with respect, and with dignity.
b. Consequences are aligned with expectations.	b. Tools, information, materials, facilities, systems, and equipment are available, accessible, labeled, usable, and accommodate the limitations of human nature and capabilities.	b. Managers and supervisors regularly reinforce safe behavior and reward accomplishments.
c. Expectations are communicated to performers.	c. Plans, procedures, job aids, and other guiding documents are usable and available.	c. Incentives and disincentives are consistent with expectations.
d. Roles, responsibilities, authorities, and priorities are specified.	d. Proper controls and barriers are in place for CRITICAL STEPS and related Risk-Important Actions (RIAs).	d. Disincentives exist for unethical, at-risk, and reckless behavior.
e. Risk importance relative to safety, quality, and production, and related CRITICAL STEPS (if any) and STOP-work criteria, are denoted.	e. Means of communication are available and workable.	e. Performance is recognizable, measurable, and traceable to the individual.
f. Plans, procedures, job aids, and other guiding documents are technically accurate.	f. Standby support, expertise, and reserves are available for unanticipated situations relative to high-risk activities.	f. Financial and non-financial incentives and disincentives are contingent on performance.
g. Relevant past operating experience is identified, reviewed, and personalized.	g. Obstacles that inhibit safe and reliable performance are eliminated, minimized, or accommodated.	g. Opportunities for success and career advancement fit with individual needs and preferences.
h. Feedback relative to performance of expectations is specific, timely, frequent, and personal.		h. The values and beliefs of the individual's immediate work group support safe practices, acknowledge human fallibility, and promote a chronic uneasiness toward the workplace.

4. Knowledge and skills

a. People understand expectations, desired accomplishments, and criteria for success.

b. People possess the knowledge and skills necessary to perform expectations.

c. People understand the key **assets** to protect, their **hazards**, Critical Steps, error traps, and the potential consequences of improper performance.

d. People retain proficiency for safety-critical activities.

e. People acknowledge unacceptable practice (e.g., cardinal rules).

f. People understand the importance of expectations to safety, quality, delivery, and production (the business).

5. Capacity and readiness

a. People are physically and mentally healthy, rested, undistracted, and fit for duty.

b. People are available for work.

c. People possess the abilities, intelligence, sociability, aptitude, size, strength, and dexterity to perform the work as expected.

d. People are able to adapt, solve problems, and think laterally for the sake of safety.

e. People agree with performance requirements, expectations, improvement opportunities, and other conditions associated with expectations.

6. Personal motives and preferences

a. People acknowledge their potential to err.

b. People possess a chronic uneasiness toward threats and **hazards** to **assets**.

c. People are willing to do what is right for safety regardless of mood or what others say or do.

d. People have positive experiences with leaders, the work, and the organization.

e. People feel that the work and expectations are meaningful and relevant to their success.

f. People are confident in their ability to perform expectations.

g. People desire existing incentives and agree with the disincentives for unacceptable practices.

h. Positive relationships with co-workers, supervisors, and managers persist, and people willingly give and accept feedback.

(Built on Gilbert, T. (1996). *Human Competence: Engineering Worth Performance* (Tribute Ed.). Washington, DC: ISPI (pp. 82–97).)

Appendix 2
Common error traps

Table A2.1 Common error traps*

1. Expectations and feedback	2. Tools, resources, and job-site conditions	3. Incentives and disincentives
a. unavailable or inaccurate procedures	a. hurrying / rushing	a. competing goals or incentives
b. weak communication protocols	b. confusing or misleading displays or controls	b. production demands of immediate supervisor
c. weak shift turnover	c. poor labeling or out-of-service instrument	c. risk-taking workgroup norms
d. unclear risk and **assets** to protect	d. distractions; interruptions; noise	d. habitual use of at-risk practices / cutting corners
e. unclear job purpose, priorities, or objectives	e. similarity of controls / no distinctions	e. boring, monotonous work
f. conflicting direction from supervisor	f. conflicting conventions / population stereotypes	f. personality conflicts with co-worker or boss
g. lack of or poor feedback from supervisor	g. complexity; high data / information flow rate	g. excessive group cohesiveness (group think)
h. simultaneous, multiple tasks	h. difficult to use or unusable procedure	h. peer pressure
i. unclear roles and responsibilities / expectations	i. adverse habitability of workplace	i. unfair, dishonest, or disrespectful boss
j. differences between expectations and norms	j. recent changes; departures from routine; workarounds	j. protracted time on job, delays, or idle time
k. lack of preparation time	k. unavailable tools, parts, help	k. distrust among co-workers / workgroups
l. overemphasis on schedule adherence	l. repetitive actions	l. lack of accountability for performance
m. meaningless rules / instructions	m. poor equipment layout; inaccessible equipment	m. progress / results are not traceable to individual
n. excessive communication requirements	n. hidden equipment response	n. punishing or fearful work conditions
	o. unexpected equipment conditions	o. "rule book" culture; cookbooking
	p. backshift; recent shift change	

4. Knowledge and skills

a. unfamiliarity with task
b. inexperience or lack of proficiency
c. lack of knowledge or skill
d. first time performing task
e. new technique not used before
f. unaware of CRITICAL STEPS and key **assets**
g. unaware of **asset's** critical parameters
h. unaware of roles, responsibilities / expectations
i. unaware of key information sources
j. invalid assumptions
k. informal communication habits
l. weak problem-solving skills
m. unaware of **hazards** or threats to key **assets**
n. tunnel vision (lack of big picture)
o. inadequate mental tracking skill

5. Capacity and readiness

a. illness, incapacity, or fatigue
b. high workload / stress
c. high concentration for prolonged periods; protracted vigilance
d. high recall (memory) requirements
e. habit patterns; recent change in environment
f. boredom
g. major life **event**
h. cognitive biases (mental shortcuts / heuristics)
i. limited short-term memory; forgetful
j. lack of manual dexterity
k. physical limitations compared with work demands
l. instinctive actions or reflex
m. post-meal slump / drowsiness
n. imprecise physical action
o. spatial disorientation
p. weak social skills
q. "loner", dislikes working with others

6. Personal motives and preferences

a. conflicting personal goals
b. production, "get-r-done" predisposition
c. overconfidence
d. willingness to bypass the rules for personal gain
e. disregard toward **hazards** to **assets**
f. argumentative and defensive
g. selfish and unethical
h. routine (perception that nothing will go wrong)
i. anxiety, nervousness, or frustration
j. fear of failure or retribution
k. excessive professional courtesy; deference to those in positions of authority
l. excessive group cohesiveness / social deference
m. no sense of control / "learned helplessness"

* Not a complete list. Categorization based on personal knowledge and experience.

Appendix 3
Fundamental human performance tools

A human performance (**Hu**) tool is a set of discrete behaviors that help workers perform their activities more reliably for specific work situations—minimizing the chances of losing control during work. These tools are considered *skills-of-the-craft*, and therefore, not proceduralized. They can be applied as needed anywhere in a work activity—every tool is designed to slow you down to give you time to think. An effective H&OP strategy depends, in part, on providing front-line workers with mental and social tools to better anticipate, monitor, respond and learn during hands-on work— RISK-BASED THINKING. Recall that the stronger **Hu** tools possess more features of RISK-BASED THINKING.

> *Warning!* Be wary of relying on **Hu** tools as your sole approach to "managing" human error. Yes, using **Hu** tools will reduce the error rate, but they will not eliminate errors completely. **Assets** still need protection (defenses) from random errors that still occur despite people's care and rigor. Too often, managers think that H&OP is simply about using **Hu** tools—and nothing could be further from the truth.

This appendix describes 10 tried and true **Hu** tools—a shopping list of sorts that applies to one or more phases of the *work execution process*. These **Hu** tools were derived from the mental and social skills used regularly by top-performing individuals and organizations. Some were adopted from military applications. Each tool's purpose is briefly described, followed by how the tool may be applied. Users are encouraged to adapt these fundamental **Hu** tools to better fit the unique constraints of their work to avoid failures for which generic tools would otherwise be ineffective. Once you decide on your "tool kit," I want to encourage you to prepare a pocket-sized booklet that can readily be used and referenced in the workplace—like a tool. The following **Hu** tools are considered public domain—feel free to use them as is:

- Pre-job briefing
- Two-minute drill
- Self-checking

- Peer-checking
- Procedure use and adherence
- Stop when unsure
- Three-part communication
- Conservative decision-making
- Turnover (handover)
- Post-job review and reporting.

Hu tool #1: pre-job briefing

Prior to beginning work, front-line workers review procedures and other guidance to familiarize themselves with what the work intends to accomplish and what to avoid. During a brief meeting, all persons involved clarify the task goals, roles, assets, intrinsic risks of the job, execution constraints, communication methods, and safety procedures. Everyone clearly understands the importance of coordination within and across the team regardless of rank. Preparation on the day of the job confirms for each individual how his or her actions affect safety and production. **RU-SAFE** adds structure to the conversations of pre-job briefings:

1 *Recognize* **assets** important to safety, quality, reliability, the environment, and production. (It may be important to protect other **assets**.)
2 *Understand* **hazards** (intrinsic sources of energy, mass, and information) to each **asset** and relevant lessons learned from previous **events**.
3 *Summarize* the Critical Steps and related RIAs.
4 *Anticipate* errors for each Critical Step and relevant error traps.
5 *Foresee* worst-case consequences for an error at each Critical Step if control of energy, mass, or information is lost.
6 *Evaluate* the controls and barriers needed at each Critical Step, the contingencies, and STOP-work criteria.

Contingencies are actions a worker intends to take in response to anticipated conditions which are known to be possible during a process but which may or may not actually come to pass. An **RU-SAFE** preview can be used while planning or preparing for work as well as during work execution. It can also be used by an individual to think through a task before a specific action is performed. This is especially beneficial when a worker comes upon a Critical Step in the workplace, not previously identified during planning or preparation.

Hu tool #2: two-minute drill

While a pre-job briefing prepares a worker mentally for a job, the two-minute drill highlights the physical reality of a situation. This tool involves taking a couple of minutes, more or less, before starting work, to become acquainted with the conditions and layout of the immediate physical workplace. The worker examines the physical surroundings to look for **assets**, **hazards**, sensitive equipment, error traps, and conditions inconsistent with the pre-job briefing. Recognizing **assets**, abnormal conditions, and troublesome **hazards** is an essential step in error-free and **event**-free performance.

This is the time to talk with co-workers or a supervisor about unanticipated **hazards** or conditions and the necessary precautions. Precautions might include eliminating **hazards**, augmenting or installing defenses, developing additional contingency plans, or revising STOP-work criteria before proceeding with the job. The individual also confirms that conditions are in line with expectations. A typical two-minute drill involves three steps:

1   *Scan* the work site, including adjacent surroundings, for one to two minutes, walking and looking around to identify **assets** and unsafe and hazardous conditions.
2   *Talk* with co-workers or the supervisor about unexpected **hazards** or conditions, especially those different from what were discussed during the pre-job briefing, and the precautions to take.
3   *Eliminate* **hazards**, minimize distractions, avoid interruptions at CRITICAL STEPS, install appropriate barriers, and develop additional contingencies before proceeding with the task.

Hu tool #3: self-checking (STAR)

Before executing a CRITICAL STEP, top performers instinctively and consistently stop and think about what they are about to do. Self-checking is likely the most common—and effective **Hu** tool. Self-checking is an attention management method that helps a performer focus on the proper component; to think about the action, the expected results, and contingencies before acting; and to verify the outcome after the action. The acronym **STAR** outlines a memorable and effective pattern for self-checking:

1   *Stop.* Pause at steps that involve a transfer of energy, information, or substance—focus attention and eliminate distractions.
2   *Think.* Consider what you are about to do to what. Is it the right thing to do? Confirm the right component with the procedure. Understand the procedure's intent and the expected result of the action. What will be done if unexpected results occur? Is there any doubt? Stop when unsure.
3   *Act.* Perform the correct action on the correct component.
4   *Review.* Verify that the anticipated result is obtained or implement a contingency action if necessary.

These steps assume the worker has the technical knowledge and skills to perform the task at hand and simply needs the additional margin for safety that self-checking brings to the job. When used rigorously, self-checking boosts attention and thinking just before performing a physical action or manipulation. Before taking an action that transfers energy, moves mass, or transmits information, the performer pauses to focus his or her attention, followed by a moment or two to reflect on what is to be accomplished—the intended action, the component, and the expected result of that action. The performer thinks about whether the proposed action is the correct action for the situation and what to do if an unexpected result occurs. If uncertain, he or she resolves any questions or concerns before proceeding. When sure of the plan, the action is performed, verifying the correct component, followed by a review of the result of the action. If the expected outcome is not evident or different than anticipated, the person stops work, placing everything in a safe condition, and either implements a contingency or gets help as needed to verify true conditions before proceeding with the work.

The Japanese enhance self-checking with a technique they call "shisa kanko," which literally means "pointing and calling."[1] The technique involves making large gestures and speaking out the status, heightening focus and attention. Japanese trained performers stretch their arms out to point at what they need to check and then name it out loud. A 2011 study by Osaka University showed that when asked to perform a simple task, workers make approximately 2.5 mistakes per 100 actions. When using shisa kanko, the number of errors dropped to near zero.[2] The pointing and calling technique has been scientifically proven to reduce error rates, improve memory performance, and in some cases, increase the speed of accurate decisions.

Hu tool #4: peer-checking

Peer-checking involves close monitoring of a performer's actions by a second knowledgeable person who is prepared to catch and prevent an error by the performer.

This tool takes advantage of a "fresh set of eyes" not encumbered by the performer's task-focused mindset. A peer may see **hazards** or potential consequences the performer does not see. Peer-checking can also be used to verify that a given plan of action or document is appropriate before further work proceeds. The worker and his or her peer (checker) agree together that the action to be performed is the correct action on the correct component. Peer-checking augments self-checking but does not replace it. A common peer-check protocol entails the following steps:

1   The *performer* and the *peer agree* on the intended action to take and on which component. Otherwise, they determine the proper action before proceeding.
2   The *peer prepares* to stop the *performer* in case an error is about to be made.
3   The *performer executes* the intended action on the correct component, while the *peer observes* to make sure the proper action is taken on the correct component.
4   The *performer* and the *peer confirm* the expected results after the action.

Front-line workers should be encouraged to ask for a peer-check when they feel the risk or conditions warrant it. Even better, a peer-check can be required by management for certain high-risk operations. A culture that values peer-checking also welcomes a peer-initiated check rather than waiting to be asked. If a person other than the performer anticipates that an action by the worker may be critical, unsafe, or otherwise, at risk—especially if a safety system is being bypassed—he or she may question the performer to verify the intent and desired result before the action is taken.

Hu tool #5: procedure use and adherence

Technical procedures direct actions for a task in a predetermined sequence and minimize reliance on a worker's memory and on ad hoc choices made in the field. A written set of instructions improves repeatability and predictability, reducing ambiguity and uncertainty of performance of complex work activities. For complex processes, procedures also play an important role in maintaining a process system's equipment configuration.

In a methodical, robust, and mindful organization, technical procedures are the outcome of a lot of expertise and are extensively tested, and will normally be accurate and up to date. Front-line workers must be committed to implementing them carefully, but mindfully. However, errors can occur in the best procedures, so front-line workers should never assume that all procedures are flawless. No procedure is a substitute for human intelligence. Remember, crossing a street at the crosswalk does not guarantee safety—you're still in the *line of fire*—the illusion of safety. Similarly, following procedures mindlessly is not inherently safe. Proper use of technical procedures adheres to the following mindful practices:

1　When working with paper copies, compare the working copy of a procedure to the control copy to verify that all revisions have been incorporated.
2　Keep the procedure in the user's presence for quick and easy reference during the work activity. If a job involves multiple work locations, ensure each location of a team has reference to or possesses a copy of all relevant procedures.
3　Review the procedure prior to starting work to verify prerequisites, that initial conditions are met, and limits and precautions (**assets** and **hazards**) are understood.
4　Use a place-keeping method to avoid missing a step, duplicating a step, or otherwise performing steps out of sequence.
5　Read and understand each step before performing it.
6　Follow directions as written in the sequence specified without deviating from the original intent and purpose unless approved otherwise by a technical authority.
7　Complete each step before starting the next step.
8　Do not mark a step "N/A" (not applicable) unless approved by a technical authority.

9 If the written instructions are not technically correct or are confusing, or the action will damage equipment if taken as prescribed, cannot be performed as written, will result in exceeding one or more of a system's or an **asset's** critical parameters, or is otherwise unsafe, then stop the task, place the process system in a safe, stable condition, and contact a supervisor (or persons with appropriate expertise).

10 If the desired or anticipated results are not achieved, stop the task, place the process system in a safe, stable condition, and contact a supervisor (or persons with appropriate expertise).

11 Review the document at the completion of the job to verify all actions were completed in the desired sequence.

Hu tool #6: stop when unsure

When doubt or uncertainty arises, the front-line worker stops work until the situation is resolved to eliminate the doubt. The use of this tool presumes the worker has checked all process instrumentation or field sources of information. The delay allows everyone involved, including the supervisor and those with relevant expertise, to resolve the issue before resuming the job. (Sometimes this tool is referred to as a "time-out.")

Never proceed in the face of uncertainty! When people are confused, uncertain, or doubtful as to how to proceed, the chances for error increase dramatically. During a personal conversation with Dr. James Reason, he said the chance for error in such situations is a toss-up, 1 in 2, "if you're lucky." Consequently, an individual is wise to stop and get help from others who have relevant expertise. *Never gamble with safety!*

"Stop when unsure" provides an overarching reminder that in any situation—whether related to a written procedure or not—in which a worker is unclear on what will happen upon taking action, he or she is obligated to **STOP** work until the issue can be clarified. *Nothing is always as it seems.* Any person can stop any activity, using the following protocol, if he or she is uncertain:

1 *Stop and secure.* Stop all transfers of energy, movements of mass, and transmissions of information, and place equipment in a safe, stable condition.
2 *Involve.* Consult co-workers and team members. Get the facts!
3 *Notify.* If still unsure, speak with your supervisor or team leader and experts.
4 *Resolve.* Solve the problem conservatively before resuming work. (See conservative decision-making in Chapter 4.)

Hu tool #7: three-part communication

Three-part communication is the formal verbal exchange of information between two people involving information related to the safety state of people, products, or property. All communications during work execution, especially at CRITICAL STEPS and their RIAs, should be formal, clear, concise, and free of ambiguity. This is especially crucial with verbal communications—whether face-to-face, via telephone, or by radio—requiring multiple exchanges. Information related to equipment or product status or personnel protection is of primary importance.

To assure mutual understanding in three-part communications essential to the success of critical operations, each participant (sender and receiver(s)) is assigned a specific responsibility in the communication. First, the sender gets the attention of the intended receiver, using the person's name. Then the sender speaks the message. The receiver repeats the message back in his or her own words. Paraphrasing by the receiver helps the sender verify that the receiver understands the intended message. Equipment names, identifiers, and data should be repeated back exactly as spoken by the sender. Finally, the sender acknowledges that the receiver heard and understood the message. Here's the three-part communication protocol in a straightforward format:

Sender "states"

1   The sender gets the attention of the receiver, such as using a first name or a position identifier.
2   The sender states the message clearly and concisely.
3   "At (location), action verb, device name."

Receiver "repeats"

1   The receiver paraphrases the message in his or her own words.
2   The receiver repeats action words and equipment designators and nomenclature as stated by the sender word for word—verbatim.
3   If necessary, the receiver asks questions to verify his or her understanding of the message.

Sender "confirms"

• If the receiver correctly understands the message, the sender states, "That is correct."
• If the receiver does not understand the message, the sender states, "That is wrong," and restates the message.

The third step in the communication is the weak link, since the sender is tempted not to pay attention to the receiver's statement, assuming the person heard the message. So, if the receiver does not understand the message, he or she should be assertive and ask the sender to repeat or paraphrase the message until definitive understanding is established. Throughout three-part communication, use a standard means of clarifying, such as the phonetic alphabet.

Hu tool #8: conservative decision-making (GRADE)

When safety is unclear for a particular action or phase of work, front-line workers, as well as their supervisors and managers, should defer to safety, not production. They should use a conservative approach every time there is any conflict between safety and production. When safety and production conflict, safety should be the default response. However, because of the complexity of operations (it's difficult to know all there is to know), conservative decision-making requires judgment under uncertainty, ambiguity, and, usually, time pressure.

Conservative decision-making in an emergency situation requires a structured, forward-looking thought process to identify and establish a safe and reliable system configuration, especially for situations that have no apparent procedure. To aid one's "thinking" in response to such uncertain, time-pressured situations, a pre-determined, structured protocol is necessary to promote conservative decisions.

CAUTION: The less time available to make an informed decision, the more readily one should default toward safety without regard to production goals.

1 *Goal.* Identify the desired safety state of the **asset(s)** in question (pinpoint critical parameter(s)).
2 *Reality.* Recognize the current safety state of the **asset** (critical parameters) and their threats (existence of **pathways**).
3 *Analyze.* Determine the difference between the goal state and current state (define the safety problem). Develop options to restore the **asset** to the goal safety state. Understand the costs, benefits, and risks of each option.
4 *Decide.* Choose and execute a course of action that will achieve the desired safety state. Decide who will do what by when. Identify key indicator(s) of the **asset's** goal safety state (critical parameters).
5 *Evaluate.* Monitor changes in the **asset's** critical parameters to verify recovery to a safe state. Review the effectiveness of the decision in achieving the goal safety state.

Hu tool #9: turnover (handover)

Turnover is the systematic exchange of work-related information between two individuals, one off-going and the other on-coming, and the subsequent formal transfer of responsibility. Usually, turnovers occur during around-the-clock operations, for the permanent transfer of responsibilities between two individuals for tasks that exceed one shift (a day's work) in length. A turnover (sometimes referred to as a handover) provides time for the on-coming individual to create an accurate mental picture of current work activities before assuming shift responsibilities or commencing work. A good turnover helps the on-coming individual understand where things stand currently and what is expected to occur during the ensuing shift.

Turnovers must be thorough and accurate, as well as brief and simple. Individuals conduct turnovers visually, verbally, and in writing. Before handing over official responsibility for the job, the off-going person should be confident that the on-coming individual is capable of assuming the duties and responsibilities of the work station and planned tasks. The following protocol is commonly used:

1   (Off-going) *Record* information, such as the following, in a turnover log:

    i   evolutions (including routine) in progress;
    ii   unscheduled work;
    iii abnormal conditions or configurations;
    iv summary of operations completed and planned.

5   (On-coming) *Review* turnover log and walk down the work location.
6   (Both) *Discuss* information face-to-face. The off-going individual ensures the on-coming operator has an accurate understanding of current status.
7   (Both) *Transfer* responsibility.

Hu tool #10: post-job review and reporting

A post-job review is learning after doing—a routine meeting conducted after work to collect feedback from the front-line worker. Sometimes called after-action reviews, this informal, brief meeting provides workers and their supervisors with an opportunity to discuss and document significant differences between work preparation (*work-as-imagined*) and work execution (*work-as-done*). Post-job reviews collect fresh insights into the weaknesses of management systems, highlighting surprises, what went well and pinpointing opportunities for improvement, personally and organizationally. If a job is important enough to have a pre-job briefing, it is important enough to have a post-job review, especially if the job included CRITICAL STEPS.

To optimize frank and open communication, rank is not observed during a meeting—everyone is on the same level. Information collected in a post-job review is isolated from performance reviews; no career should be affected as a result of what is said in a post-job review. The following protocol is a modified version of what former fighter pilot, James Murphy (author of *Flawless Execution* (2005)), used to guide combat mission debriefs in the U.S. Air Force:

1   *What did we intend to do?* Review the pre-job briefing and other preparation activities (what to *accomplish* and what to *avoid*—*work-as-imagined*).
2   *What did we actually do?* Identify what worked well (pluses) and what did not go well (deltas), including surprises during the work execution phase, especially around CRITICAL STEPS (*work-as-done*). Encourage people to tell their stories of what happened and any adjustments they made to achieve their goals.
3   *What was different and why?* Compare what actually happened to what was planned. Discuss briefly what could have caused the differences.

4　*Report it.* Report the important pluses and deltas using appropriate communication channels or methods to your supervisor.

Notes

1　Gordenker, A. (21 October 2008). Pointing and Calling: The Japan Times. Retrieved from http://code7700.com/pointing_and_calling.htm
2　Shinohara, K., Naito, H., Matsui, Y. and Hikono, Masaru (2013). The effects of "finger pointing and calling" on cognitive control processes in the task-switching paradigm. *International Journal of Industrial Ergonomics* 43, 129–136.

Appendix 4
Candidate indicators for H&OP

The purpose of this appendix is to list possible indicators that provide insights into the organization's management of H&OP. The indicators described below are simply suggestions of what might work in any industrial facility. The reader should consider these indicators as springboards for further consideration as to how they might apply to the reader's specific operational context, which should nurture the development of more informative, facility-specific metrics.

Two primary classes of indicators are provided: lagging and leading. Lagging indicators are historical—based on what has happened (past results). Leading indicators reflect current conditions and practices—an approximation of future performance based on what is happening now (behaviors). True leading indicators are statistically validated. Develop some leading indicators characterized as positive in that the indicator increases for increasing levels of safety. Leading indicators help managers monitor the presence of safety through practice and their capabilities associated with RISK-BASED THINKING.

Candidate lagging indicators

Equipment reliability traditionally is monitored by trending the mean time between failures or other forms meaningful to the organization. **Events** attributable to **Hu** can be monitored similarly. Trending the average number of days between **Hu events** indicates how well the system (blunt end) supports **Hu** at the operational level (sharp end). A **Hu event** is one that is triggered by human action (or inaction) where an **asset(s)** suffers serious harm involving the uncontrolled (1) transfer of energy, (2) movement of mass (solids, liquids, or gases), or (3) transmission of information. The failure of several defenses usually combines with the practitioner's error or violation to bring about consequences severe enough to exceed a certain severity threshold (reset criteria). Recalling that serious **events** are organizational failures, the real value of H&OP indicators and associated trending is the ability to monitor organizational performance and SYSTEMS LEARNING—the detection and correction of decline in management systems, processes, values, and operational conditions and practices.

Many work groups use the frequency of less severe **events** to help managers detect system vulnerabilities. Such criteria should be at a threshold below the site **Hu event** severity reset criteria, but strict enough to reveal hidden system vulnerabilities on a regular basis (such as every couple of months) at the work group level. Revealing work group vulnerabilities and correcting them can help minimize more serious **Hu events**.

High-severity lagging indicators

The following measures describe example indicators of various business results and work outcomes related to safety, quality, production, delivery, and cost.

1 *Site event rate.* This site-wide indicator is the 12-month rolling average number of **Hu events** per 10,000 worker-hours worked.[1] This is the primary indicator of system performance over time for **Hu events** that meet pre-determined severity criteria.
    Calculation: (number of **events** meeting facility severity criteria (over past 12 months) × 10,000) / (total worker-hours worked over past 12 months)

2 *Facility event-free clock.* This measure, based on the mean time between **Hu events**, indicates the occurrence of the most severe **events** suffered by the facility over a given period of time. It is calculated by averaging the number of days between **events** triggered by front-line personnel over last two years. A two-year rolling average is used to reduce indicator volatility and to assure the indicator reflects operational performance during extended maintenance periods. The absolute value of the calculation is not what is important to monitor—it's the trend. If the trend in the **event**-*free clock* grows (average number of days between **Hu events** trending upward), the organization is learning and becoming more resilient.

> *Caution!* Do not set a numerical value goal. The trend is what is important. Do not stop learning when the metrics look good!

Calculation: (number of days between successive facility **events** that have occurred in the last two years) / (number of **events** triggered by front-line workers).

Hu events are those **events** triggered by either active or latent errors as defined here:

- ° *Active errors*—observable, physical actions that change the state of an **asset**, resulting in immediate undesirable harm. The organization must define its **assets** and degree of harm (undesirable consequences).
- ° *Latent errors*—actions and decisions that result in organizationally related weaknesses, hidden conditions, or other flaws and defects that lie dormant until revealed by normal operation, human error, testing, or self-assessment. Latent errors that occurred within one year of the **event** are counted to be more reflective of current human performance.

3   *Work group **event**-free clock*. This measure is similar to the facility **event**-free clock except it is applied at a work-group level. The criteria for what defines an **event** are necessarily tighter, at a lower level of severity, focusing on the outcomes unique to a particular work group. The calculation is the same as for the facility **event**-free clock, the only difference being the criteria used to define an **event**.

4   *Frequency of violations or findings related to human performance*. This measures the organization's ability to find and correct its own problems before discovery by an outside agency including corporate oversight organizations.
    Calculation: (number of violations or findings) / (designated period).

5   *Ratio of severe events compared to all events*. This indicator measures the effectiveness of learning. As the organization identifies and corrects its system weaknesses contributing to weak defenses and land mines in the workplace, there will be fewer serious **events**. A lowering percentage indicates learning is effective.

Caution! Do not set a numerical value goal. The trend is what is important. Do not stop learning when the metrics look good!

Calculation: (number of **events** meeting a given severity level) / (number of all **events** regardless of severity).

Low-severity lagging indicators

The following situations are generally associated with work outputs as denoted in the *systems thinking model*. Conceivably, you could make the case that these indicators are leading. Although these situations are arguably "after the fact," they are minor conditions or practices that precede more serious outcomes, which, if monitored closely, could provide time for managers to take proactive, systemic corrections to avoid a serious **event**. Another way to view the following situations is that such minor occurrences can adversely influence *work-as-done* to deviate from *work-as-planned*. Again, it is suggested that these situations be used as springboards for inspiration and creativity to develop site- or group-specific indicators.

- Incorrect assembly of a component.
- Minor injury or first aid.
- Not attending scheduled training.
- Missing or worn-out tools.
- Near-hit error.
- Inaccurate procedure, checklist, or operator aid.
- Unfamiliarity with high-risk work.
- Damage or deterioration of important system components.

- Equipment unavailability or non-conformity.
- Automation malfunction or related software errors.
- Miscommunication during high-risk work.
- Fatigue, illness, or drug-related impairment (for people performing high-risk work).
- Missing or inaccurate quality certifications.
- Delay or change in scheduled work.
- Security breach or unauthorized access to facility premises or information.
- Excessive consecutive number of hours worked or lack of sufficient number of hours of sleep (by day or week).
- More than two concurrent activities (when one task is considered high-risk work).
- Recurring deviations from specific good laboratory practices or good manufacturing practices.
- Problematic work activities (those experiencing recurring errors, mishaps, or **events**).
- Housekeeping infractions in the workplace.
- Mispositioned operational components.
- Pre-job briefings not conducted for high-risk work.
- Insufficient or untimely resources.
- Mis-labeled components or items used in operations.
- Procedure revisions not made before being used again.
- Minor spills, leaks, or transfers.
- Work-arounds (workplace conditions requiring recurring compensating operator action).
- Inadequate or missing supervision for high-risk work.
- Recurring false or locked-in annunciator alarms.
- Misleading or incorrect data (used to make operational decisions).
- Postponing or canceling scheduled training, observations, audits, or inspections.
- Contamination occurrences.
- Schedule conflict for high-risk work or multiple high-risk jobs scheduled for same place at same time.
- Signal passed at danger (actions contrary to safety signals).
- Recurring equipment issues or problematic equipment, instruments, or tools.
- Workplace deficiencies not corrected promptly.
- Persistent error traps for high-risk work (target those most worrisome to the work for resolution).

Candidate leading indicators

Leading indicators provide managers with the ability to make proactive interventions more quickly and avoid serious harm. Preferably, indicators should increase as safety is improved instead of tracking to zero—positive leading indicators. To help managers identify such drivers, the selection of safety performance indicators should be soundly based on an underlying but validated safety model. Models are always simplified representations of reality. However, they highlight the most essential characteristics of a phenomenon and reveal its most relevant functions. To this end, six building blocks of H&OP are used to categorize leading indicators.[2] The building blocks model helps managers better understand "how" their systems influence safety and resilience.

Warning! As stated in Chapter 7, don't weary yourself with every "uptick" in metrics. All leading indicators typically drift up and down over time, and an organization can burn out trying to respond to all unfavorable trends. The goal should be to identify when there has been a significant change in indicators—when the underlying process has changed. This could be done by setting a threshold based on historic data (during the initial year of data gathering) and, then, counting how many times the threshold is breached afterward. Such a trend would serve as a basis for exploration as to why the change.[3]

---

1   RISK-BASED THINKING (adaptability to changing workplace risks):

- Frequency of job stoppages by workers due to uncertainty or confusion.
- Frequency of surprises encountered during operations.
- Percentage of unplanned work (emergent) compared with planned work (total number of emergent items divided by the total of all work started per period).
- Number and duration of equipment out of service.
- Differences between time scheduled for work vs. time actually started.
- Number and duration of workplace workarounds (long-term deficiencies or temporary modifications requiring operational adjustments by practitioners in the field).
- Number and duration of temporary modifications.
- Frequency of lack of resources for high-risk activities (something missing).
- Frequency of differences between *work as planned* vs. *work as done* (number of reports per period per work group).
- Overtime and excess straight time compared with total number of employees per period by work group.
- Turnover rate by work group (number of persons joining and leaving work group compared with the total number of people in the work group per period); 12-month rolling average.
- Frequency of equipment label requests (for existing equipment).
- Bench strength of expertise in critical skill areas.

2   CRITICAL STEPS (control of high-risk work activities in the workplace):

- Percentage of high-risk work plans and technical procedures that denote CRITICAL STEPS.
- Fraction of high-risk jobs that conduct pre-job briefings.
- Number of CRITICAL STEPS discovered during work execution (not previously identified) per period.
- Percentage of CRITICAL STEPS encountered that were previously identified in a procedure or pre-job briefing.
- Number of surprise CRITICAL STEPS identified (per period by work group).
- Number of RIAs not denoted in procedures or maintenance work packages.

3   SYSTEMS LEARNING (continuous improvement of managements systems and related defenses):

- Fraction of pre-job briefings that discuss relevant operating experience for task per period per work group.
- Fraction of high-risk jobs that conduct post-job reviews.
- Identification of system weaknesses, land mines, and error traps to be corrected.
- Number (or percentage) of preventive maintenance activities that reveal failure or hidden faults.
- Number of deficiencies identified by outside agencies (third parties) compared with all deficiencies identified (per period by organization).
- Number of self-assessments conducted per period per work group.
- Frequency of failures of emergency equipment.
- Rate of increase of number of deferred repairs.
- Fraction of **event** analyses that address H&OP from a systemic perspective vs. a person perspective.
- Age of open CA / PA (corrective action / preventive action) items by priority (backlog) or average time to close an H&OP corrective action.
- Fraction of system-oriented CA/PAs compared with all types of CA / PAs.
- Ratio of CA / PAs involving training or procedures revisions to all corrective actions.
- Total number of CA / PA extensions granted over the quarter.
- Procedure revision backlog (total open procedure revision requests on the last calendar day of the quarter) by administrative and technical procedures.
- Average age of technical procedure revision backlog.
- Percentage of occurrence of effectiveness reviews for CA/PAs for serious **events**.
- Number of system-level self-assessments per period.
- Fraction of **events** with repeat corrective actions.

4   *Training and expertise* (development of technical knowledge, skill, and proficiency):

- ◦ Number of job stoppages by practitioners due to uncertainty per period per work group.
- ◦ Number of qualifications per period.
- ◦ Number of hours of ongoing technical refresher training per period per work group.
- ◦ Percentage of staff who have received H&OP-related training each year.
- ◦ Timeliness of training on changes to technical procedures and equipment.
- ◦ Fraction of workers assigned to attend training who actually attend.

5   *Observation and feedback* (managers seeing firsthand how work is done, while promoting two-way learning):

- ◦ Time (hours per week) devoted to observations of work by senior managers and executives.
- ◦ Fraction of work activities that report issues.
- ◦ Frequency of self-reporting by work group.
- ◦ Percentage of corrective actions involving disciplinary action or counseling.
- ◦ Fraction of all reports that are self-reported (per period by work group).
- ◦ Number of whistle-blower reports to outside regulatory agencies (per period by work group).

6   *Integration and execution* (enabling risk-based thinking throughout the organization and managing change: plan–do–check–adjust):

- ◦ Percentage of high-risk operations and procedures that have identified CRITICAL STEPS and related RIAs.
- ◦ Frequency of job stoppages due to lack of resources.
- ◦ Resources devoted to training and management of H&OP.
- ◦ Frequency of visits by CEO or VPs to production sites.
- ◦ Frequency of visits to shop floor by senior managers and executives.
- ◦ Frequency of H&OP-related topics on the agendas of Board of Directors, management review meetings, and other senior management meetings.
- ◦ Frequency of meetings that address H&OP or Risk-Based Thinking concurrently with operational issues.
- ◦ Budget for staff augmentation for high-tempo activities (beyond current capacity).

Notes

1 Alternatively, to be consistent with the Occupational Safety and Health Administration (OSHA) you can use 200,000 hours in the formula, which represents the equivalent of 100 employees working 40 hours per week, 50 weeks per year. (See U.S. Bureau of Labor website: https://www.bls.gov/iif/osheval.htm.) Choose a normalization value that is meaningful to the organization.

2 Wreathall, J. (2006). Properties of Resilient Organizations: An Initial View. In Hollnagel, E., Woods, D. and Leveson, N. (Eds.) *Resilience Engineering: Concepts and Precepts*. Aldershot: Ashgate (pp. 279–280).

3 This warning was suggested by John Wreathall through personal correspondence.

Appendix 5
The building blocks of H&OP— self-assessment criteria

An assessment provides a standardized, disciplined approach to determining the effectiveness of current practices compared with best in class, and identifying areas for improvement. This appendix describes the more important success factors associated with each building block of Human and Organizational Performance. An initial assessment helps identify the starting point for executing H&OP and integrating RISK-BASED THINKING into your operations.

1   *Adapt*—RISK-BASED THINKING. Intent: enhance the workforce's capacity to adapt to changing risks. As an operating philosophy, RISK-BASED THINKING augments the capacity of front-line workers to recognize changing operational risks and to adjust their behaviors appropriately. These abilities are enhanced first by a humble acceptance of one's own human fallibility, and second by an intentional pattern of thought: anticipate, monitor, respond, and learn. Additionally, people reflect a deep-rooted respect for the technology and its intrinsic **hazards** as well as an on-going mindfulness of impending transfers of energy, movements of mass, and transmissions of information.

   a. The following core practices of RISK-BASED THINKING (habits of thought) are integrated into operational activities:

   ○ *Anticipate*—means to foresee a) the implications of the task at hand and its intended accomplishments, b) the **assets** and intrinsic **hazards** involved, and c) the potential consequences to **assets** to avoid if a loss of control of **hazards** occurs during work.
   ○ *Monitor*—means to recognize a) the creation of **pathways** between intrinsic **hazards** and key **assets**; b) impending transfers of energy, movements of mass, and transmissions of information (i.e., CRITICAL STEPS); and c) undesirable trends of an **asset's** critical parameters that define its boundaries for safety (or onset of harm) or for quality.
   ○ *Respond*—the mental and physical capabilities to exercise positive control of intrinsically hazardous operations to a) control human actions, b) protect **assets** from the onset of harm, or c) limit the degree

of harm, using resources that are available in the workplace, here and now.

  ° *Learn*—the means to know a) how control was lost and how harm ensued in previous **events** similar to the task at hand, b) in real time what is happening to **assets** and **hazards**, and c) what must be changed going forward to ensure the integrity of defenses (controls, barriers, and safeguards).

b. People do not proceed with a high-risk task if uncertain or surprised. *Never proceed in the face of uncertainty* until the safety of **assets** can be assured!

c. All operational decisions conservatively protect key **assets** when production goals conflict with the safety of those **assets**. When the unexpected occurs, people with the greatest expertise, regardless of position, are involved in the decision-making process.

d. People, regardless of position, freely challenge assumptions, insist on facts, and consider potential consequences before proceeding. When uncertain, they seek the counsel of those with the requisite technical expertise.

e. Managers and other leaders actively promote a chronic uneasiness toward work (transfers of energy, movements of mass, and transmissions of information) by talking about it often in a variety of formal and informal settings.

f. Managers inform front-line workers that they can take prompt action they are qualified to render anytime to protect themselves, others, and key **assets** from harm.

g. Everyone avoids the notion that any high-risk activity is "routine," and regularly ask "What if…?"

2 *Control*—CRITICAL STEPS. Intent: identify and control **Hu** variability at CRITICAL STEPS. By necessity, the organization must be sensitive to operations—where transfers of energy, movements of mass, and transmissions of information regularly and frequently occur. Controlling the performance at CRITICAL STEPS improves first-time quality—a means of "source inspection," and tends to decrease the frequency of **events**.

a. Define what constitutes a CRITICAL STEP for each work group, if necessary. For consistency, reserve the phrase "critical step," for work activities that involve operational, hands-on work.

b. CRITICAL STEPS, related RIAs and their controls are identified during the work-planning and procedure development processes. CRITICAL STEP MAPPING may be helpful in accomplishing this initially.

c. Before starting work, front-line workers acknowledge the important **assets** involved in the assigned tasks and are mindful of their intrinsic

hazards. They know what they expect to accomplish, when **pathways** exist, but are wary of their fallibility.

d. Pre-job briefings and post-job reviews are systematically conducted for high-risk work, including those with one or more CRITICAL STEPS.

e. Front-line workers understand and follow technical procedures (until proven unusable or unsafe) and use **Hu** tools, mindfully, especially where it counts for safety, quality, and reliability.

f. Workers avoid at-risk and unsafe practices, and so-called "necessary violations" to get work done, especially at CRITICAL STEPS and RIAs.

g. Regardless of the absence of any **hazards**, CRITICAL STEPS are *always* considered critical—pulling the trigger on a firearm is always critical.

4   *Learn*—SYSTEMS LEARNING. Intent: discover and correct weaknesses with defenses. **Events** are organizational failures, and the causes of tomorrow's **events** exist today. Therefore, it is incumbent on all managers to be relentless in the identification and correction of system-level weaknesses and vulnerabilities that tend to diminish the integrity and robustness of built-in defenses or inhibit the organization's resilience. When effective, SYSTEMS LEARNING tends to reduce the severity of the few **events** that still occur.

a. Managers are accountable for the identification and correction of system weaknesses, land mines, and faulty or missing defenses in their areas of responsibility.

b. Changes to facilities and management systems avoid creating volatile (uncontrollable) situations or reducing flexibility and options for the workforce.[1]

c. Managers follow through with improvements promptly, not postponing needed changes.

d. The management team regularly reviews the performance of the organization's various management systems and organizational functions, aided with metrics that look at both lagging and leading indicators.

e. Line managers are skeptical of indicators trending positively and positive reports. They assume the worst for declining indicators and negative reports. No news is assumed to be bad news.

f. Managers conduct regular self-assessments of their management systems, work practices, and work conditions.

g. Managers benchmark high-performing/high-reliability organizations to enhance the organization's resilience against threats to safety, quality, and reliability.

h. Managers systematically take advantage of lessons learned from past work and industry operating experience to improve **Hu** and prevent **events**.

i. Managers use a corrective action / preventive action (CA/PA) process that supports regular identification, prioritization, analysis, and timely resolution of problems and opportunities.

j. Managers are aware of the extent of condition of issues—avoiding simply treating symptoms.

k. Managers analyze **events** using a systems perspective (i.e., systems thinking). "Near hits" (close calls) are treated as **events**.

l. Managers encourage investigations of **events** assuming people want to do a good job and that people had logical reasons for doing what they did in an **event**.

m. The senior manager bonus structure reinforces system-wide safety as well as shareholder return. Relevant indicators of major **hazard** risks are incorporated into the bonus systems and performance agreements of executives and site or division managers, who have the capacity to influence corporate or group outcomes.[2]

4   *Training and expertise.* Intent: develop and sustain technical expertise. Technical training is an essential element of H&OP. In particular, technical expertise is the bedrock of RISK-BASED THINKING and the recognition and control of CRITICAL STEPS. Using a systematic approach to training ensures people understand the organization's business, possess a thorough understanding of the operation's technology, and have a deep-rooted respect for its intrinsic **hazards**.

a. Line-training activities optimize technical expertise in front-line workers, incorporating H&OP principles, practices, and vocabulary. Include scenario-based opportunities to practice RISK-BASED THINKING, chronic uneasiness, and mindful use of selected **Hu** tools.

b. People are trained and qualified before doing work unsupervised. Front-line workers possess the technical knowledge and skills required for the job. Also, they are given a broad range of experiences to develop their depth of knowledge of not only the technology but also how their work fits into the organization's business and the purposes of management systems relevant to their work.

c. Develop and conduct training using a systematic approach, such as ADDIE (*A*nalyze needs, *D*esign the training, *D*evelop training material, *I*mplement training, and *E*valuate the training effectiveness).

d. Provide training that informs people of their capabilities and limitations as human beings as well as a means to control the occurrence of human error, especially during CRITICAL STEPS. Additionally, training emphasizes RISK-BASED THINKING, chronic uneasiness, conservative decision-making, and learning from mistakes.

e. Conduct refresher training regularly to strengthen technical knowledge and skills, informing front-line workers of changes in the technology, as well as lessons learned from **events**. People should be familiar with operations or functions outside their own areas of expertise.

f. Train front-line workers on how to use and adapt **Hu** tools, giving them opportunities to practice using **Hu** tools in actual or simulated work settings.

g. Resources for technical training and refresher training are preserved even when the company experiences periods of austerity.

h. Executives and senior managers ensure adequate staffing bench strength in critical positions.[3]

5   *Observation and feedback*. Intent: improve manager awareness of what is actually happening in the workplace. Managers regularly and frequently monitor front-line operations, engaging front-line workers face-to-face to remain aware of what is actually happening in real time, giving and receiving feedback.

   a. Executives and line managers spend time on the shop floor regularly to stay aware of current **Hu** issues in the workplace.
   b. Managers and first-line supervisors are expected to observe work. Line managers pay close attention to day-to-day operations, especially during high-risk, high-tempo periods.
   c. The criteria for a "good" observation is defined.
   d. The occurrence of observations by line managers is tracted on an individual basis.
   e. Managers define expectations—what is acceptable and unacceptable for high-risk work, in particular, **Hu** tools. Managers clearly communicate expectations to front-line workers and are able to model them. **Hu** tools are adapted to suit the work's unique circumstances.
   f. Exercise accountability at all organizational levels, while treating people with dignity and respect.
   g. Give people appropriate feedback:

      ◦ *Reinforce* people when they exhibit safe practices.
      ◦ *Coach* people when their practices are inconsistent with expectations.
      ◦ *Correct* people who exhibit unsafe, reckless, unethical acts, or make recurring at-risk choices.
      ◦ *Console* people when their errors trigger harm.

   h. Managers and supervisors do not tolerate reckless, unethical, and recurring at-risk choices (drift).
   i. Regularly encourage employees to report problems, regardless of significance, especially recurring differences between *work-as-done* and *work-as-planned*. Actively solicit information from front-line workers on ways to improve safety, quality, reliability, and production.
   j. Front-line workers willingly give and accept feedback. They do not ignore unsafe, reckless, unethical, or at-risk practices by co-workers.

6   *Integration and execution*. Intent: the systematic and disciplined management of **Hu** risks, while instilling Risk-Based Thinking into all facets of operations. Integration *enables* the cornerstones of Risk-Based Thinking in the various organizational functions and operational processes. Execution involves managing the operational risks of **Hu** and a systematic follow-through to *align* the organization to sustain all of H&OP's building blocks.

   a. Executives and line managers consider H&OP as core business—a management responsibility, and their decisions and actions are aligned with the principles of managing H&OP.

b. Executives issue a written policy on the management of H&OP, consistent with current orders, policies, codes, standards, practices, and regulations. Managing H&OP is not optional.

c. Executives commit the necessary resources to safety and resilience regardless of the absence of **events** or the presence of economic pressures. Resources do not ebb and flow with normal fluctuations in indicators.

d. Line managers have knowledge of and respond to important operational **Hu** and related system issues.

e. Executives and line managers communicate the importance and value of managing the risks of human error in operations regularly and frequently.

f. Line managers align operational processes, resources, and planning to support the safety and resilience of high-risk work activities.

g. Leaders purposefully and accurately use the H&OP vocabulary—frequently referring to RISK-BASED THINKING and its cornerstone habits of thought.

h. RISK-BASED THINKING is promoted more as a way of thinking and doing business than as a program or a checklist.

i. RISK-BASED THINKING habits of thought are integrated into various organizational functions, management systems, and key operational processes—enabling *anticipate, monitor, respond,* and *learn.*

j. Line managers align *organizational* and *local factors* to enable and sustain expectations and to inhibit unwanted practices.

k. Front-line workers readily perceive the economic limitations of their work—they understand the organization's business and mission.

Notes

1 Roe, E. and Schulman, P. (2008). *High Reliability Management: Operating on the Edge.* Stanford, CA: Stanford University Press (p.213).

2 Hopkins, A. and Maslen, S. (2015). *Risky Rewards.* Farnham: Ashgate (pp.146–159).

3 Critical positions are ones that could affect adversely key performance measures such as safety, revenue, costs, quality, or customer-engagement indicators, or are critically important to the business strategy. An absence of qualified workers for a critical position can have devastating results on the organization.

Glossary of terms

If you want to change the way you think, change the words you use.

Karl Weick

If one is to effectively manage the risks associated with human error during operations, a vocabulary must exist to talk and communicate about it. Definitions are important to managing **Hu** in the workplace—*you cannot manage what you do not understand*. A vocabulary improves your understanding of a domain of knowledge, aiding your ability to manage it. Therefore, I've included this glossary. It will serve as a ready reference as you read this book. I encourage you to flag this page as you will likely refer to this section often during your readings.

Accident   An unplanned, surprise occurrence that results in injury, damage, or loss, usually occurring within moments of a trigger. (Compare with event.)

Accomplishment   A product of work behavior that delivers value toward achieving the organization's mission; work output. (See work.)

Accountability   The expectation that an individual or an organization chooses to be answerable—willingly takes responsibility—for the outcomes of their choices (given the circumstances at the time). (Compare with local rationality.)

Accumulation   The unnoticed and unabated buildup of at-risk or unsafe conditions in an organization or workplace over time. Examples: latent system weaknesses, faulty or missing defenses, land mines, error traps, or unsafe beliefs or values. (See land mine, latent condition, and latent system weakness; compare with drift.)

Action   Observable, physical bodily movement that involves the application of force. (See active error and behavior; compare with error and touchpoint.)

Active error   An error (action or inaction) that unintentionally triggers a loss of control of a hazard, altering the state of an asset, resulting in its immediate harm. (See action and error; compare with latent error and violation.)

Alignment   The extent to which the organization's system achieves its business goals through the influence of people's behaviors. The arrangement of local factors to influence specific behavior choices required to produce work outputs that accomplish the organization's business purposes. (See systems thinking.)

Anticipate To know what to expect for the work-at-hand—to foresee the implications of a proposed act or course of actions. Includes looking ahead for possible pitfalls, and potential consequences to assets, given the hazards and their pathways, exemplified by asking "What if...?" (See chronic uneasiness and Risk-Based Thinking.)

Asset Something of high value to an organization that could sustain injury, damage, or loss and is important to an organization's mission, survival, and sustainability.

Examples:

- people (health and well-being—the most important asset);
- product (quality);
- property (facilities, tools, and equipment);
- environment (earth, water, and air);
- time (productivity and schedule);
- public trust and reputation;
- money (cash, capital, margin);
- proprietary information;
- software;
- intellectual property.

At-risk choice A behavior choice when a person consciously acknowledges there is risk but believes he/she is still safe; the person either doesn't comprehend the potential harm that could occur or erroneously feels the risk is justified under the circumstances.[1] (Compare with just culture and recklessness.)

Attitude Feelings (favorable or unfavorable) a person has toward an object, subject, task, or another person.

Barrier Means used to protect an object (or asset) from harm by limiting or impeding uncontrolled transfers of energy (such as gravity, electrical, chemical, heat, kinetic), movements of objects or substances (such as loads, shipments, product, fluids), or transmissions of information (such as signals, digital records, software). (See control, safeguard, and defense.)

Behavior The mental and physical efforts to perform a function; observable (movement or speech) and hidden (perceiving, thinking, analyzing, deciding) tasks or activities by an individual. (See action and work.)

Benchmarking A process of comparing products, processes, and practices against exemplars, the best in class, the toughest competitors, or those companies recognized as industry leaders. Provokes discovery, innovation, new perspectives, and possibly dissonance with current processes and practices.

Blunt end The upper echelons of an organization—those who are removed in time and space from direct production activities, who organize and control its activities, set policies and priorities, manage business processes, decide on technical design, provide resources, and prepare procedures. People who influence (directly and indirectly) the performance of production work of front-line workers. (Compare with sharp end.)

Business case   The reasoning, usually documented, to assess the benefits, costs, risks, and options for a proposed new and significant activity. Reasoning for a proposed course of action that is aligned with the organization's purposes and values.

Chronic uneasiness   An ongoing mindfulness about one's work exhibited by:

- a preoccupation with impending or current transfers of energy, movements of mass, or transmissions of information that could threaten the safety of an asset;
- an awareness of the presence of hidden hazards in the workplace, such as land mines and error traps;
- a sensitivity to one's fallibility—the capacity to err (requires humility);
- a reluctance to consider any job or situation as "routine."

(See anticipate and Risk-Based Thinking.)

Coaching   Feedback to an individual or small workgroup through face-to-face dialogue that facilitates change in a person's behavior related to a specific expectation. (Compare with correction and reinforcement.)

Complacency   Personal satisfaction with the safety or security of a current situation, unaware of actual dangers, hazards, defects, or deficiencies. Being unconcerned while performing work in a hazardous environment. (See chronic uneasiness.)

Complex   An operational situation characterized by the presence of many, concurrent, and obscure interconnections, unusual conditions, and individually adaptive components (such as people), that can include unanticipated domino (ripple) effects, not easily comprehensible to any one person.

Condition   Something that is fixed, or exists unchanging (at a point in time)—a circumstance, situation, factor, that exists when taking an action. (See latent condition.)

Examples:

- fluid characteristics: temperature, density, flowrate, etc.;
- wellness of persons, such as health, strength, and fatigue;
- distance of an object above the ground or floor;
- weight, speed, acceleration of an object;
- labels: placement and legibility;
- presence and integrity of barriers, such as guardrails, fences, locked doors, interlocks, passwords, shielding, insulation, etc.;
- knowledge, experience, and proficiency of persons;
- an open or closed door, latched or unlatched;
- a burned out (blown) light bulb;
- equipment in working order, such as full functionality of operating and safety components of a circuit breaker.

Note that some conditions compared with others can be temporary—existing momentarily—such as weather, staffing levels, software glitches, the absence of supervision, etc.

Conservative decision-making   A decision that places greater value on an asset's safety than on achieving the immediate production goal—resulting in taking actions to place the asset in a known safe condition. The intent to preserve or improve an asset's margin of safety. (See trade-off.)

Control Means that guide, coordinate, or regulate behavior—promoting desired actions and outcomes, while improving the chances of error-free performance Sometimes used as a synonym for a defense, as in an engineered "control." (See barrier, defense, and positive control.)

Correction Feedback to an individual intended to stop an unethical, at-risk, unsafe, or reckless behavior. (See feedback; compare with coaching and reinforcement.)

Critical Step A human action that will trigger immediate, irreversible, intolerable harm to an asset, if that action or a preceding action is performed improperly. (See Risk-Important Action.)
Examples:
- pulling the trigger of a firearm (whether loaded or not);
- skydiving—leaping out of the open door of a serviceable aircraft at altitude;
- closing an electric circuit breaker;
- clicking "send" or "submit" while on an internet website;
- depressing the "enter" key
- opening or closing an isolation valve.

Culture An organization's system of commonly held values, priorities, beliefs, and assumptions that influence an individual's reactions and behavior choices in the performance of their work. It includes the assumptions one has about what one must do to be effective, successful, or to survive in a situation. (Compare with attitude and chronic uneasiness.)

Danger An asset's exposure or proximity (without adequate protection) to one or more hazards that can cause harm to the asset. (See hazard and pathway.)

Defense Means taken to protect an asset against harm. Methods used to control or reduce an asset's exposure to a hazard, such as controls, barriers, and safeguards. Anything that tends to reduce the frequency or severity of events. (See control, barrier, and safeguard; compare with pathway.)

Defense-in-depth The overlapping capacity of redundant defenses to protect assets from danger, such that a failure of one defense is compensated for by other defenses, avoiding harm. (Compare with resilience.)

Drift The progressive behavioral deviation between expectation (work-as-imagined) and actual performance (work-as-done)—often in the direction of greater risk (reduce margins for safety or error). (See accumulation.)

Energy The ability to do work, e.g., heat, mechanical, gravitational (potential), kinetic, chemical, electrical, and nuclear. It is the uncontrolled release of energy that is typically very dangerous to humans and other assets.

Engineered control Those devices or structures (hardware, software, or equipment) built into the physical environment that protect assets from harm, and that function without activation by or dependence on people's actions. (See defense.)

Error A behavior that unintentionally deviates from a preferred behavior for a given situation. An act, an assertion, or a belief that unintentionally deviates from what is correct, right, or true.[2] Sometimes error involves a loss of control, but usually is not recognized as an error by those involved until after an unwanted consequence is realized. (See active error, human error, and latent error.)

Error trap　An unfavorable condition in the workplace or in the mind of the performer that either creates uncertainty or otherwise enhances the chances of error during performance of a task in the here and now. (Compare with local factors.)

Event　An unwanted occurrence involving harm (injury, damage, or loss) to one or more assets due to an uncontrolled (1) transfer of energy, (2) movement of mass, or (3) transmission of information. (Compare with accident.)

Expectation　A written statement of a desired accomplishment (work output), specifying the criteria for its success, and/or the desired behavior for a given situation.

Feedback　Information about past or present behavior, results, or conditions that gives an individual or an organization an opportunity to change. (See coaching, correction, and reinforcement.)

Harm　Injury or damage to an asset, or loss of value, capability, or impairment of mission—tangible or intangible. An unwanted change in the key characteristics or state of an asset.

Hazard　A source of energy, mass, or information that could harm an asset or cause its loss—usually intrinsic (built-in) to a facility for operational and business purposes. Any condition in the workplace that could harm or trigger harm to an asset. (See danger; compare with pathway.)

High-reliability organization (HRO)　An organization that successfully operates in an unforgiving physical, social, or political environment, rich with the potential for error, and where complex processes are used to manage a complex technology in order to ensure success and avoid failure.[3] Sometimes HRO is used to denote the processes of high-reliability organizing. (Compare with resilience.)

Human and organizational performance (H&OP)　The collective performance of an organizational unit involving the behavior choices and work outputs of several workers performing within the systemic context of the organization's technical and social environments. (Compare with human performance.)

Human error　An error that refers to the slips, lapses, fumbles, and mistakes of the human kind, regardless of whether one's goal is accomplished or not—not all human errors have bad outcomes. (See active error, error, and latent error.)

Human nature　The innate characteristics of being human (including our fallibility).

Human performance (Hu)　The behavior (B) of an individual to accomplish a specific result (R) ($Hu \rightarrow B + R$). (Compare with human and organizational performance.)

Human performance tools (Hu tools)　Distinct mental and social skills that complement an individual's technical skills to achieve safe and efficient task performance, carving out time for thinking before doing. Sometimes referred to as non-technical skills. (See control.)

Human reliability The probability of successful performance of human activities, whether for a specific act or in general—usually expressed as a percentage.

Job A combination of tasks and duties (responsibilities) that define a particular position within an organization. (See work; compare with accomplishment.)

Just culture A workplace climate that (1) assumes people want to do a good job, (2) does not tolerate reckless, unethical, and at-risk choices, (3) does not sanction human nature, and (4) encourages an organization's members to willingly report mistakes, mis-steps, and system weaknesses without fear of punishment. (See at-risk choice and culture.)

Knowledge worker An individual who works primarily with information, or one who creates and communicates knowledge for use by others. (Compare with worker.)

Land mine A metaphor for an undetected hazardous condition in the workplace that is poised to trigger harm. A workplace condition that increases the potential for an uncontrolled transfer of energy, mass, or information with one action—an unexpected source of harm—unbeknownst to the performer. Includes compromised or missing defenses; also known as an "accident waiting to happen." (See Critical Step; compare with latent system weakness.)

Latent condition An unknown situation or circumstance established before the present that persists, whether in the management system, production system, or in the workplace, which may lie dormant for long periods of time, doing no apparent harm. (See condition and latent system weakness.)

Latent error An error (action, inaction, or decision) that creates a potentially unsafe condition, unnoticed at the time, causing no immediate, apparent harm to an asset, but which could combine with other errors, occurrences, or conditions at a later time to realize harm. Typically, they manifest themselves as degradations in defense mechanisms, weaknesses in processes, land mines, and inefficiencies and undesirable changes in values and norms. (See error, latent condition, and latent system weakness; compare with active error.)

Latent system weakness Latent conditions (undetected deficiencies or defects) in management systems, facility or equipment design, work processes, plans, values, or leadership practices. Sometimes referred to as "latent organizational weakness." (Compare with latent condition and latent error.)

Learn A change in behavior in response to the acquisition of new knowledge and understanding of its meaning and application. In the context of Risk-Based Thinking, knowing requires active learning to understand:

1 what has happened (previous experiences relative to the task at hand);
2 what is happening (situation awareness of the task at hand); and
3 what to change (identification of behavior and system changes needed to enhance safety of assets and resilience of the organization for future similar operations).

(See Risk-Based Thinking.)

Line of fire Physical path that energy or an object would travel if stored energy is released, controlled or uncontrolled; an asset's exposure to harm while in the path.

Local factors A set of workplace conditions that influence the behaviors choices of the individual, while performing a task or taking an action in the here and now.[4] Sometimes referred to as "performance shaping factors." (See error trap.) (See Appendix 2.)

Local rationality The presumption that people do things that make sense to them at the time (given their circumstances)—otherwise, they would not do them.[5] The influence of local context (local factors) on decisions. (See local factors; compare with accountability.)

Loss The reduction in the value of an asset (usually monetarily), or the physical deprivation of an asset.

Management system A formally established and documented set of activities designed to produce specific results in a predictable, repeatable manner.

Mental model A person's structured thought or knowledge about how something works—a mental picture or other representation of the underlying way in which something functions.

Mistake An error committed when the intent of an action was incorrect for the situation; also, an incorrect decision or interpretation. (See error.)

Monitor Attention devoted to a particular thought or object—especially the safety of assets and their related intrinsic hazards, comparing the actual status with the desired status. In the context of Risk-Based Thinking, concentration on varying levels of risk of harm to assets such as:

- The emergence of pathways for transfers of energy, movements of mass, or transmissions of information, especially for important assets.
- Critical Steps and related risk-important actions for the work at hand.
- Important changes in an asset's key safety parameters to allow timely response to avoid harm if needed.
- Trends in key indicators of safety and resilience (leading and lagging).

(See anticipate, learn, respond, and Risk-Based Thinking.)

Motive One's personal goals, needs, interests, or purposes that tend to stimulate action in order to achieve or meet the goals.

Near hit An occurrence that under slightly different circumstances could have resulted in injury to people, damage to property, product, or environment, or other loss, but it did not occur due to chance (luck) or a good catch by an individual. More conventionally referred to as a "close call" or "near miss." (Compare with event.)

Organization A system, comprised of people, resources, and technology, designed to coordinate and direct a group's individual and collective behaviors toward the accomplishment of its mission.

Organizational factors Conditions, processes, and practices in the upper echelons of an organization, that create and moderate workplace conditions that, in turn, influence people's behavior during work.

Examples include:

- management and leadership practices;
- management systems;
- procedure development and revision;
- training programs;
- operational work processes;
- espoused values and priorities;
- structures, systems, and components of technology
- type and availability of resources.

(See local factors, blunt end, and sharp end.)

Pathway An operational situation where an asset's transformation (change in state) is poised to occur by either a transfer of energy, a movement of mass, or a transmission of information. Exposure of an asset to the potential for harm where only one action (equipment or human) is needed to alter the asset's state, for good or for bad. (Compare with line of fire.)

Positive control A person's attempt to ensure that what is intended to happen is what happens, and that is all that happens. (Compare with action.)

Process A series of activities or functions organized into tasks to produce a product or service. Means established to direct the behavior of individuals in a predictable, repeatable fashion as they perform various tasks.

Recklessness The conscious disregard for taking a substantial and unjustifiable risk—total unconcern for the potential consequences of an action, without intending to cause harm. (Compare with at-risk choice and violation.)

Reinforcement Feedback that an individual experiences, through the use of positive consequences, when a specific behavior occurs—increasing the likelihood that the behavior will occur again. (See coaching and feedback.)

Reliability The assurance that a function (human or machine) is performed to a certain standard when it is called upon. "The lack of unwanted, unanticipated, and unexplainable variance."[6] (Compare with resilience.)

Resilience The intrinsic ability of a system to adjust its functioning before, during, and after a challenge, disturbance, or failure to sustain operations under both expected and unexpected conditions.[7] The ability of a system or process to return to a desired and acceptable state following an upset with little to no harm to its assets. The ability to succeed under varying circumstances—including the capacity of front-line workers to adapt to changing risk conditions in the workplace. (Compare with defense-in-depth, high-reliability organization, reliability, and safety.)

Respond Knowing what to do (1) to protect assets from harm, (2) to recover from a harmful situation, or (3) to create a better outcome. In the context of Risk-Based Thinking, response involves enhancing the front-line workers' capacity to adapt to changing risk conditions—especially during surprise situations—to protect assets from harm, such as:

- Exercising positive control of Critical Steps (whether planned or unplanned).
- Building slack or buffers into operations or tasks through resources, time, alternatives, flexibility, expertise, contingencies, etc.

- Taking actions to either (1) eliminate the task or operation, (2) prevent error, (3) catch an error before it triggers harm, (4) detect a defect, or (5) mitigate harm done.

(See anticipate, chronic uneasiness, learn, monitor, positive control, and Risk-Based Thinking.)

Risk   The likelihood of an asset suffering an unwanted consequence (injury, damage, or loss); any occasion an asset is exposed (opportunity) to a threat (hazard). (See pathway, Risk-Based Thinking and severity.)

Risk-Based Thinking   Four interdependent habits of thought (anticipate, monitor, respond, and learn) in the conduct of work that enhances a person's readiness and the capacity to adapt by building work-specific knowledge about the safety of assets involved in the work activity.

Risk-important action (RIA)   A reversible human action preceding a Critical Step that either:

- creates a pathway for the transfer of energy, movement of mass, or transmission of information that expose an asset to a hazard (for the conduct of work);
- influences the number or effectiveness of defenses that protect assets (barriers and safeguards);
- impacts the ability to maintain positive control of the release of energy, mass, or information (hazards) at Critical Steps;
- verifies conditions important for safe operation at Critical Steps (to avoid land mines).

(See Critical Step.)

Root cause   Those systemic factors (organizational) in an event such that had they not been present before the event, the event would not have happened. Alternatively, those systemic factors preceding an event, which, if corrected, will prevent recurrence of the event. Note: Systemic corrective actions are necessary to resolve root causes. Systemic interventions change the fundamental way a system works to achieve its desired organizational and operational functions. (See systemic.)

Safeguard   Means of mitigating or minimizing the harm done to an asset after the onset of injury, damage, or loss. (See defense; compare with barrier and control.)

Safety   An asset's freedom from unacceptable risk of harm.

Safety culture   An environment where people at all levels of the organization make safe choices that protect assets from harm. The strength of a safety culture is measured by the proportion of employees who steer away from at-risk choices.

Self-assessment   Formal or informal processes of identifying one's vulnerabilities and opportunities for improvement by evaluating present practices, conditions, and results against desired goals, industry best practices, policies, expectations, and standards. (See benchmarking.)

Severity   The degree of harm an asset suffers (or could have suffered), such as: none, minor, major, disastrous, catastrophic (defined differently by different organizations). Usually the amount of harm realized is governed by the intensity or magnitude of the energy transferred, the amount of mass moved, the sensitivity of the information transmitted, and the durability of the asset at the time of the event.

Sharp end Individuals in an organization who are in direct contact with production assets, safety-critical processes, and intrinsic hazards, who perform hands-on work to achieve the organization's purposes. (See worker; compare with blunt end.)

Situation awareness The accuracy of a person's current knowledge and understanding of working conditions compared with actual conditions at a given time.

Skill-of-the-craft A set of knowledge and skills related to certain aspects of a task or job that an individual is expected to know and be fluent in without needing written or verbal guidance.

Supervisor A member of management who directs and oversees the performance of one or more individual contributors (front-line workers) in the conduct of assigned production activities. Sometimes referred to as a "first-line manager."

System A collection of elements or components arranged in a particular way that function together in an environment, responding to inputs to produce predictable, repeatable outcomes (most of the time). Complex systems sometimes operate in unanticipated ways producing surprising outcomes.

Systemic A characteristic of a system's structure or operation that has the power to influence the choices of all individuals in that system at the same time—in the here and now, instead of individual members or parts (not to be confused with "systematic," which means methodical). (See root cause.)

Systems thinking An understanding of how systems behave and how their components interact with other components (and with other systems) to influence (1) people's choices and their outcomes in the workplace, and (2) the resilience (defenses) of hazardous operational processes.

Task A planned work activity with distinct start and stop points made up of a series of intervening actions of one or more people to accomplish one or more objectives, usually directed by a procedure or skill-of-the-craft. Sometimes used to denote a single, distinct action. (See skill-of-the-craft; compare with action.)

Touchpoint A human interaction with an object (including assets and hazards) that changes the state of that object through work. Work usually initiates a transfer of energy, a movement of mass (solid, liquid, or gas), or a transmission of information. (Compare with Critical Step.)

Trade-off A choice between being efficient (or productive) and being thorough (or safe)—accepting less of something you value to get more of another thing you value. It is rarely possible to be both completely efficient and completely thorough at the same time. (See conservative decision-making.)

Uncertainty Being unsure of what to do—a feeling of doubt. One's ambiguity about the current or future states of an object or situation, usually based on insufficient knowledge or the complexity of the situation.

Values Central principles and/or possessions held in high esteem by the members of the organization around which decisions are made and actions occur. (See asset and culture.)

Violation A behavior that intentionally deviates from an expected behavior, sometimes for purposes of a personal nature. (See motive; compare with error.)

Work The application of a force over a distance, usually for the purpose of adding value. (See drift and accomplishment; compare with energy.)

Workaround A deficiency in a procedure, component, or workplace that compels a front-line worker to take an adaptive, compensatory action—usually manual—different from the guiding procedure, to achieve one or more goals. (See drift.)

Work-as-done Work as it is actually accomplished by workers in the workplace. (See accomplishment; compare with work-as-imagined.)

Work-as-imagined Work as envisioned, planned, or believed to be happening by managers, designers, and procedure writers. Sometimes known alternatively as "work-as-planned." (Compare with work-as-done.)

Worker An individual in the organization who performs physical, hands-on work, having direct contact with (capable of altering the condition of) equipment, assets, intrinsic hazards, and safety critical processes. Also, referred to as a front-line worker. (See sharp end; compare with knowledge worker.)

Work execution The process of accomplishing production activities that involve (1) preparation for work, (2) execution of work, and (3) learning from the work accomplished. (See work.)

Workplace The physical location (local) where people can touch and alter assets and equipment, or interact with intrinsic hazards and safety-critical processes. (See local factors and worker.)

Notes

1 Marx, D. (2009). *Whack-a-Mole: The Price We Pay for Expecting Perfection.* Plano, TX: By Your Side Studios (pp.33–37).

2 American Heritage Dictionary. Retrieved from http://www.yourdictionary.com/error#americanheritage. Retrieved 2 May 2017.

3 Rochlin, G.I. (1993). Defining "high reliability" organizations in practice: A taxonomic prologue. In Roberts, K. (ed), *New Challenges to Understanding Organizations.* New York: Macmillan (pp. 11–32).

4 Reason, J. (1998). *Managing the Risks of Organizational Accidents.* Aldershot: Ashgate. (p.11).

5 Dekker, S. (2014). *The Field Guide to Understanding 'Human Error'* (3rd edn). Farnham: Ashgate (p.6).

6 Hollnagel, E. (1993). The Reliability of Interactive Systems: Simulation-Based Assessment. in Wise, J., Hopkin, V.D. and Stager, P. (Eds.) *Verification and Validation of Complex Systems: Human Factors Issues.* Berlin Heidelberg: Springer-Verlag (p.209).

7 Hollnagel, E. (2012). *How do we recognize resilience?* Presentation at University of BC School of Population and Public Health Learning Lab., 7 May 2012.

Further reading

I recommend reading the following titles for those who are interested in furthering their education in Human and Organizational Performance. Executives and line managers new to H&OP should first read those titles listed under the Beginner heading, and then explore titles under the Intermediate heading that are of interest to you. Of those on the Beginner list, I suggest you first read Dr. Reason's book, *Managing the Risks of Organizational Accidents* followed closely by Dr. Dekker's *The Field Guide to Understanding 'Human Error'*. David Marx's novel, *Dave's Subs*, about accountability describes in simple, direct terms what a "just culture" looks like and how to manage it. Advanced titles tend to be more academic in nature, offering less immediate practical applications. However, I encourage human performance specialists to study as many titles as possible of all difficulty levels. And, I strongly suggest all line managers eventually read Dr. Hollnagel's book, *Safety-I and Safety-II*, which looks at the future of safety management.

Beginner

- *Managing the Risks of Organizational Accidents*, James Reason
- *Field Guide to Understanding 'Human Error'* (3rd edn), Sidney Dekker
- *Dave's Subs: A Novel Story About Workplace Accountability*, David Marx
- *Pre-Accident Investigation*, Todd Conklin
- *Managing the Unexpected: Resilient Performance in an Age of Uncertainty* (3rd edn), Karl E. Weick and Kathleen Sutcliffe
- *A Life in Error*, James Reason
- *Safe by Accident?*, Judy Agnew and Aubrey Daniels
- *The Industrial Operator's Handbook*, H.C. Howlett
- *Flawless Execution*, James Murphy

Intermediate

- *The Human Contribution*, James Reason
- *Occupational Risk Control*, Derek Viner
- *Barriers and Accident Prevention*, Erik Hollnagel

- *Exemplary Performance*, Paul Elliot and Alfred Folsom
- *The ETTO Principle: Efficiency-Thoroughness Trade-Off*, Erik Hollnagel
- *The High-Velocity Edge*, Steven Spear
- *The Observant Eye*, W.T. Subalusky
- *Safety at the Sharp End: A Guide to Non-Technical Skills*, Rhona Flin, Paul O'Connor, and Margaret Crichton
- *Good to Great* (for business leaders), Jim Collins
- *The Checklist Manifesto: How to Get Things Rights*, Atul Gawande
- *To Engineer is Human*, Henry Petroski
- *The Design of Everyday Things* (2nd edn), Don Norman
- *The Luck Factor*, Richard Wiseman

Advanced

- *Human Competence: Engineering Worthy Performance*, Thomas Gilbert
- *Safety-I and Safety-II*, Erik Hollnagel
- *Behind Human Error* (2nd edn), David Woods et al.
- *Human Error*, James Reason
- *Engineering a Safer World*, Nancy Leveson
- *Error in Organizations*, David Hofmann and Michael Frese (Eds.)
- *Guidelines for Preventing Human Error in Process Safety*, American Institute of Chemical Engineers
- *FRAM: Functional Resonance Analysis Method*, Erik Hollnagel
- *Drift into Failure*, Sidney Dekker
- *High Reliability Management: Operating on the Edge*, Emery Roe and Paul Schulman
- *Safety Differently*, Sidney Dekker
- *Normal Accidents: Living with High-Risk Technologies* (2nd edn), Charles Perrow

Index

Page numbers followed by *n* indicate note numbers.

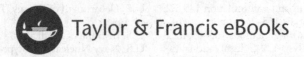

Printed in the United States
by Baker & Taylor Publisher Services